Practical Poultry Feeding

Practical Poultry Feeding

RAY FELTWELL and SYD FOX

FABER AND FABER
London & Boston

First published in 1978
by Faber and Faber Limited
3 Queen Square WC1N 3AU
Printed in Great Britain by
Latimer Trend & Company Ltd Plymouth

British Library Cataloguing in Publication Data

Feltwell, Ray
Practical poultry feeding.
1. Poultry—Feeding and feeds
I. Title II. Fox, Syd
636.5′08′4 SF494

ISBN 0–571–10817–2
ISBN 0–571–11231–5 Pbk

Contents

Illustrations

Acknowledgements

The authors are most grateful to the following for their co-operation:

Agricultural Research Council, London, UK
Beecham Research International, Brentford, UK
Dr Norman Comben, Roche Products Ltd, Dunstable, UK
Feedstuffs, The Miller Publishing Company, Minneapolis, USA
Dr John Feltwell, Sutton Valence School, Kent, UK
Dr A. William Jasper, The American Farm Bureau, Illinois, USA
H. J. S. Jones, Plumpton Agricultural College, Sussex, UK
Dr N. R. Knowles, The British Egg Marketing Board, Research Education Trust, Tunbridge Wells, UK
Dr Philip Lee, Spiller International, London, UK
David Martin, Hubbard Farms, Stroud, UK
John Pearson, Simon Barron, Gloucester, UK
Dr Peter Pilbrow, Pilwood Feeds Ltd, Hampshire, UK
L. J. Scott, Colborn International, Kent, UK

The photographs have very kindly been supplied by:
Dr Norman Comben, Roche Products Ltd, Dunstable, Uk; plate 1
Dr C. A. Elvehjem, University of Wisconsin, USA; plates 2, 3
Dr Paul Hilbrich, Hanover, West Germany; plate 4
C. R. Grau, University of California, USA; plate 5
H. Kivimoe, Uppsala, Sweden; plate 6
Butyl Products Ltd, Billericay, UK; plate 7
Pertwee Landforce Ltd, Colchester, UK; plate 8
Christy and Norris Ltd, Chelmsford, UK; plate 9
E. and B. Equipment Ltd, Barnsley, UK; plates 10, 11, 12, 13

Our thanks are due also to Richard Bird who has provided an index of great value.

Preface

The 'revolution' in poultry husbandry has been one of the more significant phenomena in food production in the last quarter-century. The comparative efficiency with which chickens and turkeys turn vegetable material into protein-rich foods for human consumption has prompted the application of much advanced technology in already advanced poultry industries and also the establishment of similar industries in many areas where poultry keeping had never been more than a domestic issue and a rather neglected one at that.

As a result poultry keepers representing a very wide range of both experience and inexperience are continually seeking information to help them improve productivity in their flocks. For those whose production methods are already firmly based on what the biological sciences have to offer, several more-detailed scientific texts on poultry nutrition are available. It seemed to the authors, however, that a more simplified, mechanistic approach to poultry feeding would be attractive both to them and to many whose scientific background might be less than fully established. In this book the science of nutrition is summarized only to draw from its complexity a platform on which sound practical poultry feeding can develop.

The bulk of the book is taken up with the problems of obtaining satisfactory rations for one's birds, either by knowing exactly what service the poultry-food manufacturer ought to be offering, or by taking over his role, on a small or perhaps larger scale, where, as in many countries, his services are not yet established. For this reason we have devoted considerable attention to the processes of milling and to the possible usefulness of the many unfamiliar foodstuffs as well as those that are standard components of modern rations. We hope, however, that what will be the most interesting and useful contribution of the book, is the detailed attempt to explain the matching of formulations to the production standards to be expected of the various classes of stock.

Ray Feltwell,
Hartfield, Sussex

Syd Fox,
The University,
Reading

CHAPTER ONE

Poultry nutrition – the process and the factory

Metabolism in the chicken

Metabolism is the continuous process by which food ingredients are absorbed and utilized by the body. More strictly, it is a combination of two simultaneous processes of predominantly chemical change whereby, on the one hand, feedstuffs and some stored tissue, such as body fat, are broken down into simpler chemicals and, on the other, new and complex molecules are again built up to form the various tissues of the body and its products. These processes are respectively those of (a) catabolism, which yields energy for warmth and activity as the larger molecules are broken down and (b) anabolism, which is the process of construction.

In contrast with some other animals, domestic fowls have a rapid digestive process. From the time when the chicken has eaten its food to the time when it voids the residues, a period of not more than twelve hours (and perhaps as little as four hours) may have elapsed. It is, therefore, very important to know the factors on which the efficiency of the chicken's digestive system depends in order to maintain that efficiency and to regain it when the process breaks down. Food costs represent 65 to 75 per cent of the total costs of a poultry enterprise, and good or faulty feeding together with proper or improper food use by the chicken are the factors likely to make the largest impact on the success or failure of a poultry project.

Accordingly, the aim of this section is to review the nutritional process of the chicken from a mechanistic and practical point of view. All the emphasis is placed upon an examination of possible faults that may develop and upon the ways in which they may be cured. To this end the digestive system of the chicken is described here rather like a factory, and digestion as the chemical and physical processes which the factory incorporates in order to achieve a desired product.

Both parts of the subject are treated in the same chapter since although it is possible to admire both the complexity and the compactness of the lay-out of the factory, no real judgement of its technical efficiency (in either biological or economic terms) can be made until its operation can be understood against the back-

Table 1 Summary of processes involved in poultry nutrition

PROCESS	INPUT	PRIMARY AND INTERMEDIATE METABOLIC PROCESSES		OUTPUT
Description	Provision of feedstuffs	Conversion of raw feedstuffs into chemical compounds meeting the nutritional needs for: (i) Maintenance	(ii) Production of eggs or table meat	Growth Eggs Fertility
Examples	(a) Cereal grains (b) Extracted vegetable seed meals (c) Vitamins and minerals	(a) Oxidation and hydrolysis of feedstuffs (b) Maintenance of body temperature (c) Maintenance of muscular and respiratory activity and all other normal bodily functions	(a) Organization of metabolic pathways to provide anabolic products making up each end product	
Important considerations bearing on these activities	(a) Cost (b) Physical form of ration (c) Balance of ingredients (d) Cleanliness and freedom from spoilage, taints and moulds	(a) Environmental temperature and shelter (b) Feather cover	(a) Health status of birds	(a) Quality of product in relation to: (i) incubation and hatching (ii) human consumption (b) Price structure of the market

ground of the process that it has to accomplish. As in any industrialized system, this biological process has its inputs and its outputs and if we measure these accurately we can compare one against the other for a measure of efficiency. Indeed we attempt just this when we calculate 'kilograms' or 'pounds' of feed consumed per kilo of eggs produced or per kilo of liveweight gain, and when we include the resultant figure in our economic forecasts of the likely profit or loss of our poultry enterprise.

Of course, such calculations of profit or loss are governed by variations in the cost of feedstuffs and in the market price of eggs and table chickens and by the way these variations synchronize or fail to match. But it is the efficiency of the 'on-line' process which determines what can be made of a favourable price situation and what can be done to offset a bad one.

Nor, unfortunately, is it sufficient merely to balance the final output against the input if we are concerned to exercise control over possible variations in the efficiency of the system. This is because food utilization for egg or meat production is not a single-stage process. We can summarize the important issues as in Table 1.

Summary of the processes of digestion and absorption

Only a brief summary of these complex processes is given here. For a detailed treatment the reader should refer to the texts listed in the bibliography.

Poultry feeds, whether the scraps of the farmyard or balanced mixtures purchased from a merchant, are made up of animal and vegetable products containing all the main groups of nutrients: carbohydrates, fats, proteins, minerals and vitamins. Chemically speaking they are very complex substances which must be digested; i.e. they must be broken down into simple, water-soluble and diffusable products so that they can be absorbed into the bloodstream through the mucous membrane of the intestine.

The process of digestion may be regarded as being partly mechanical but mainly chemical. The mechanical side involves the swallowing of the food and its passage through the digestive system aided by peristaltic muscular action. The chemical side is dependent upon the action of secretions known collectively as the intestinal or digestive juices. These secretions contain substances known as enzymes which are biological catalysts produced by living tissues and have the ability to encourage chemical change, e.g. digestion, without themselves being changed in the process. Each enzyme has specific requirements concerning the temperature and the degree of acidity or alkalinity under which it will best work.

The intestinal juices are secreted by glands in the digestive tract itself or by organs closely associated with digestion like the liver and the pancreas.

On being consumed, food passes down the gullet and becomes moist from the action of the salivary juices; then it may either remain for a period in the crop or it may completely bypass the crop on its way to the stomach. In the stomach the food becomes permeated by the intestinal or gastric juices which contain hydrochloric acid in addition to enzymes.

The breaking down of complex nutrients is accomplished by a process of hydrolysis when the food becomes intimately mixed with the digestive juices, with bacteria and, perhaps, other microorganisms.

The enzymes involved in the breakdown of carbohydrates are known as amylases. The first amylytic enzyme encountered by food on being consumed is ptyalin which is present in the saliva, but the effect of saliva on carbohydrate breakdown is limited. In the small intestine, however, carbohydrates are turned into simple sugars (e.g. glucose), known as monosaccharides, by the action of the pancreatic enzyme amylopsin and the enzymes invertase, maltase and lactase which are present in the *succus entericus*, the mixture of juices secreted by the glands in the intestines.

The enzymes involved in the breakdown of protein are known as proteases. The digestion of protein commences with the action of the enzyme pepsin and is continued in the small intestine by the action of the pancreatic enzyme trypsin and the enzyme erepsin of the *succus entericus*, leading to the eventual release of the protein's constituent amino acids.

Enzymes concerned in the breakdown of fats are called lipases.

In the small intestine the bile salts emulsify the fats, thus allowing the lipases or steapsin of the pancreatic juices to break down the fats into their constituent fatty acids and glycerol.

Once the process of catabolism has been completed, some of the soluble nutrients, including the minerals, pass through the microscopic hair-like villi which line the walls of the small intestine and are then absorbed in one of two ways. The bulk of the nutrients, including the simple sugars and amino acids, pass directly into the hepatic portal vein which carries these nutrients to the liver. The fatty constituents, after re-forming into fats, take an indirect route to the blood through the lacteal ducts of the lymphatic system whilst the bile salts, which have played their role, also return to the liver by way of the hepatic portal vein.

The indigestible portion of the food plus the unused nutrients are voided as the droppings.

The organs of digestion – the factory lay-out

The digestive tract of the fowl is illustrated in Fig. 1 and more diagramatically in Fig. 2 where it has been divided into sections for separate study.

The tract is comparatively long and from beak to vent may be five or six times the external length of the bird (see Table 2). This fact emphasizes both the convoluted nature of the tract itself and, also, that the rapid passage of food through this length of system means that the food is present for only a short period in any one organ – when the contribution of that organ has to be made to the digestive process. The proper functioning of each part of the tract is, therefore, of great importance.

Table 2 Length of digestive tract, broiler male, 10 weeks old

Mouth and throat	5 cm
Oesophagus	31 cm
Proventriculus	6 cm
Small intestine	188 cm
Large intestine to the vent	15 cm
Total length	245 cm

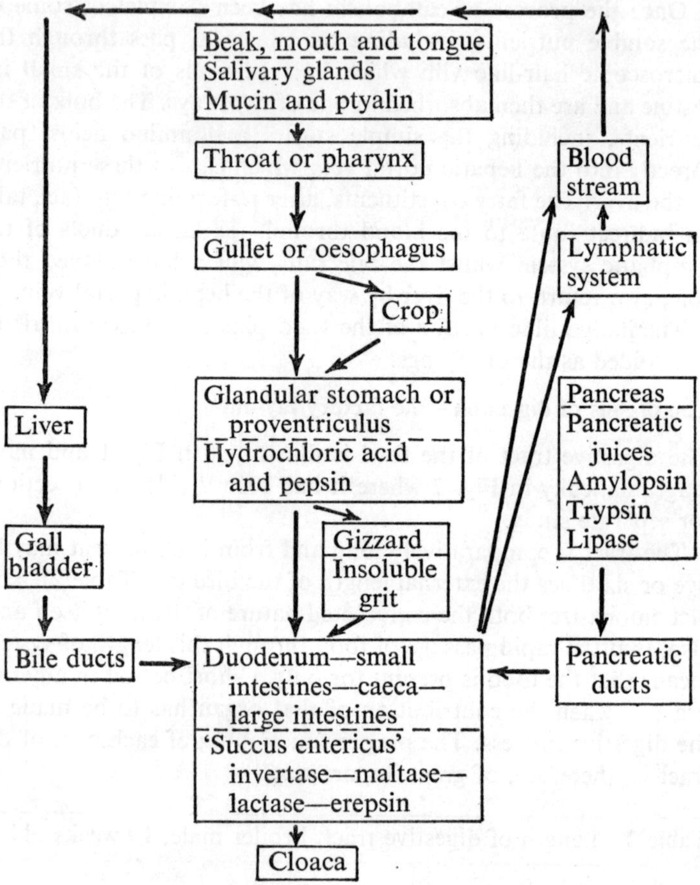

Fig. 1 Digestion: the factory lay-out

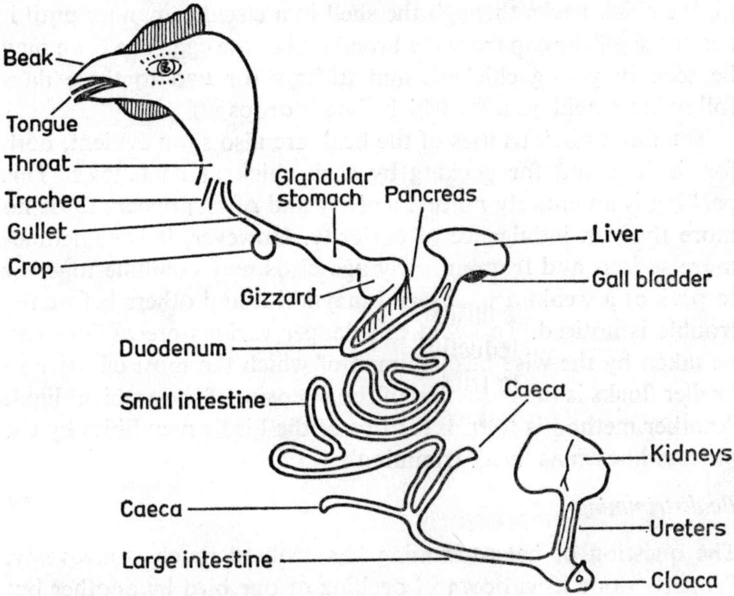

Fig. 2 Diagrammatic representation of the digestive tract

RECEPTION

The mouth

The mouth of the chicken is primarily for grasping and receiving food, and, occasionally, for use as a weapon. It does not contain teeth and so mastication has to be carried out elsewhere – in that very muscular portion of the stomach known as the gizzard. Instead, the mouth has been modified by the horny growth of the upper and lower mandibles that form the beak (in the case of chickens and turkeys) or bill (ducks and geese) so that it is well adapted to picking up small objects – e.g. grains, grits – from a flat surface.

The beak is one of the first organs of the bird to be brought into use. During incubation, the tip has an additional point, known as the bean or egg tooth, which is used in the task of breaking out from the shell. Having obtained its initial supply of oxygen for lung respiration, by piercing the air cavity within the shell, the piping chick pecks through the shell in a circular manner until it can force off the cap from the broad end of the egg. The bean may be seen in young chickens and turkeys for two to three days following hatching, after which time it drops off.

The more obvious uses of the beak are also soon evident, both for feeding and for pecking by each chick at its fellows. This pecking is an entirely normal activity and often appears to be no more than an indulgence of curiosity. However, it is sometimes more serious and from an early age birds may combine together to peck at a weakling chick and may kill it and others before the trouble is noticed. To avoid this danger various precautions can be taken by the wise husbandman of which the most effective in broiler flocks is the reduction of the intensity of the ambient light. Another method is to trim the tips of the bird's mandibles by the process known as 'beak-trimming'.

Beak-trimming

The question of beak-trimming has aroused much controversy. Properly done it will control pecking of one bird by another but will be likely to slow down food intake and hence growth rate for some time after the operation. Badly performed it will lead to considerable stress, failure to eat and reduced performance. Tender scar tissue may form where the beak is cut and may extend the period of time in which the bird is reluctant to eat. Infection and blood poisoning may occur.

Bad trimming is likely to result from the faulty use of poorly designed equipment, or from a failure to understand the physiological structure of the beak and to appreciate that the soft and sensitive tissues underlying the horny covering may not easily recover from severe cutting. Almost certainly in the past poultrymen have trimmed too severely in an attempt to prevent regrowth of the mandibles and, perhaps, avoid the need to trim again later.

In fact the beak is a complex structure with several layers of tissue of different cell types with abundant nerve and blood supplies. Where these tissues are exposed by severe trimming (perhaps two-thirds of the beak being removed) they remain sore and tender and are irritated by the meal which the bird is given to eat and by the hardness of the metal triggering device which may operate the water drinker.

Trimming is usually done with an electrically heated knife which should be hot enough and sharp enough to make a clean wound and at the same time cauterize it at a controlled temperature. A modern machine will also have a guide bar retaining the beak so that an excessive length cannot be cut from it.

Under these circumstances there is no danger in beak-trimming, but certain other conditions will minimize stress. The bird should have access to clean water so that she can wash her beak free from food and the food should be deep enough (5 cm or more) in the trough so that there is no likelihood of her banging her beak against the base. The bird should be stimulated to return to regular feeding by any means possible.

An initial fall in growth rate is, however, very likely and broilers could be expected to be 50–70 g lighter at normal slaughter age. For broilers, therefore, it is standard not to trim the beaks but to maintain the flock under conditions of low light intensity in which they are much less likely to peck around at objects other than their food. Fig. 3 illustrates various methods of beak-trimming.

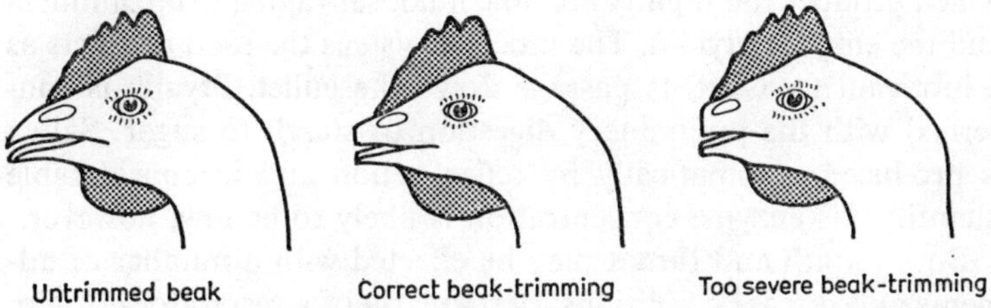

Fig. 3 Beak-trimming

Deformities of the beak

Sometimes chicks have deformed beaks, perhaps with one twisted across the other, or maybe with the lower one 'undershot' back from the level of the upper one. In the latter case it would be possible to help the situation by trimming the upper mandible. Birds with twisted beaks can hardly be expected to thrive and should be culled.

Where birds are fed a meal of too fine a texture, their beaks may become clogged or pasted. This can seriously affect the birds and may occur particularly if they are suffering from a mucous discharge from the nostril or mouth. The trouble can be remedied easily enough in the case of the individual bird but is a serious nuisance if many birds are affected.

The tongue, pharynx and gullet

The characteristic pointed tongue of the fowl assists in passing the food picked up by the mouth directly into the gullet. It is also, however, liable to get pasty with food which has been too finely ground. Mixed with saliva the food may harden into a ball which lodges underneath the tongue and may cause substantial swelling and inflammation of the skin below it. Again it is easy enough to take out the obstruction from the bird's mouth. Some birds accumulate pellets of this kind one after another.

The throat or pharynx connects the mouth to the gullet or oesophagus which runs down the line of the neck, past the crop, until it joins the glandular stomach.

The mouth and throat are well provided with salivary glands which produce the slightly alkaline fluid, saliva, made up of mucin and the enzyme ptyalin. The mucin moistens the food and acts as a lubricant to assist its passage down the gullet. Ptyalin is concerned with the preliminary digestion of starch to sugar. Saliva is produced automatically by reflex action and in considerable quantity. Its enzyme concentration is likely to be low, however.

Both mouth and throat may be effected with a number of ailments and disease conditions, particularly of a respiratory origin. Soreness will almost always cause a reduction of food intake until the illness is treated successfully with antibiotics.

Lesions may also be present in the throat due to a deficiency of vitamin A and almost complete blockage of the throat can occur with the oesophageal form of fowl pox, when birds can be seen to be avid for food but are unable to pass any into their crops.

STORAGE – THE CROP

The crop is a pendulous sack-like organ attached to the oesophagus which acts as a storehouse and thereby allows the bird to ingest a comparatively large amount of food at a time and, if necessary, go without further feeding for several hours. Thus just before darkness falls (naturally or artificially) birds can be seen eating rapidly in order to fill their crops with food to last the next three to four hours. The encouragement of this four-hour cycle of eating and digestion is one of the means of maximizing broiler growth.

The length of time food remains in the crop depends on the nutritive status of the bird and on the nature of the food. If the gizzard has become empty, new food may virtually by-pass the crop completely. Generally, however, bulky fibrous foods may remain in the crop from five to eight hours and whole grains even longer.

Softening of the food, by mixing with saliva and water, takes place in the crop and the process can be thought of as a partial pre-digestion of the food which is structural or physical rather than chemical, but which facilitates what happens later in the glandular stomach. The early moistening of the food is probably necessary for subsequent digestion and it is known that wet mash passes faster from the crop than an equivalent weight of dry meal. This perhaps is one reason why wet mash has a traditional value for increasing appetite when this has been reduced, often by illness.

Entry of food into the crop is controlled by a ring of sphincter muscles at the neck. The sieve-like ability of the crop to isolate coarser food for special treatment is sometimes disturbed either because the bird develops a depraved appetite for some fibrous material, or because grass or lucerne is eaten in such large quantities that a residue of its fibres becomes compacted and then swells in the crop as the mass absorbs water. The bird is then said to be 'crop-bound' and will require time and perhaps a dose of olive (or some other vegetable) oil, to help clear the obstruction.

Fibre may be troublesome in other ways. If it is sharp, like the lignin from the stem of post-flowering lucerne, it may actually rupture the crop and lead to internal poisoning or peritonitis.

The term 'sour crop' is used to describe one in which undesirable fermentation has taken place following the ingestion of mouldy or decaying food. It is often caused by the fungus *Candida albicans* which is particularly resistant to the antibiotics normally given to control this specific ailment. The crop contents may be forced out of the bird's mouth by gently squeezing the crop with the bird held head downwards. Spores of the *Candida* fungus may be a main cause of enteritis.

'Pendulous crops' occur, probably as a result of weak musculature at the attachment to the oesophagus. It is unlikely to be a fatal affliction but may prevent normal crop function.

The crop may also be attacked by crop worms but this rarely occurs in Europe.

PRE-DIGESTION – THE STOMACH

After leaving the crop, food passes down the oesophagus to the stomach which is divided into two portions, the glandular stomach (proventriculus) and the gizzard.

The proventriculus

The glandular stomach is concerned with chemical digestion. It is very thick walled and well supplied with glands secreting digestive juices. These juices become closely mixed with food and contain hydrochloric acid and the enzyme pepsin, whose function is the breakdown of dietary proteins.

Impaction of the glandular stomach sometimes occurs from an accumulation of grass or straw fibres.

The gizzard

The second portion of the stomach, the gizzard, is very muscular and functions almost solely as a grinding machine for physical reduction of the food-particle size. The grinding of the food is achieved by the muscles of the gizzard wall forcing food particles into abrasive contact with insoluble mineral grit, flint or granite, which the bird ingests for just this purpose. If the gizzard is empty of grit, a substantial portion of the food, especially whole grain, will be voided undigested. It follows therefore that grit of suitable particle size should be included in the diet as a standard practice.

Gizzard worms are sometimes a problem with ducks and geese. The worms are reddish in appearance and about 2·5 cm long and are found under the mucous membrane lining the gizzard.

Insoluble grit

Insoluble grit, flint and granite, helps the bird to derive the maximum value from the food supplied. Calcium grit, e.g. limestone grit, is a soluble grit and is not suitable for this grinding or milling purpose. It is often said that insoluble grit acts as the teeth of a bird and it is particularly important where whole or coarsely ground cereal grains are fed. When an all-mash diet is supplied the need for insoluble grit is apparently less absolute, but it will still give increased food utilization. There is less evidence to justify the use of insoluble grit in those cases where a high-efficiency broiler diet is used.

The size of grit depends upon the age of the bird. Chick grit is used during the first week, followed by medium-size grit up to ten weeks of age, after which adult-size grit should be used. Grit size is regulated in relation to the size of screening mesh through which it will pass but there is no standard nomenclature.

Grit is best offered by sprinkling a little on the mash during the first few days, then by supplying, at two-weekly intervals, 0·5 kg of grit per 1,000 birds. Grit is retained in the gizzard until it is worn smooth and is of smaller size when it passes on down the alimentary tract. It is possible, therefore, that an unrationed supply could result in excessive consumption. Whenever growing birds have been without grit for some time, only a little should be provided when it is reintroduced into the diet, otherwise the birds may gorge on the grit and then be unable to eat normally.

Birds on free range will frequently obtain a good deal of insoluble grit by foraging. Some turkey producers supply birds over sixteen weeks of age with a pea-size washed gravel. This is readily consumed by turkeys and apparently satisfies their requirements for insoluble grit.

Contrary to what one may at first assume, the presence of these abrasive particles in the gizzard appears to maintain the tone and strength of the muscles making up its lining. Abnormal softening and sometimes thickening of the lining – giving rise to haemorrhages and cellular erosion – seem to be associated with the feeding of very finely ground food. The feeding of coarsely ground meals and insoluble grits prevents this problem. Choline and vitamin B_{12} may also have an important role to play in this condition.

Sometimes the gizzard becomes impacted and there is an accompanying apparent failure in mobility of the muscular walls after the bird has ingested an excess of long fibrous materials. This seldom happens if the grit supply is adequate and if the diet is formulated to provide enough bulk so that the birds need for volume in its intake does not have to be satisfied by grass fibres, floor litter or feathers.

ABSORPTION – THE INTESTINES

The intestines receive the food in a creamy semi-solid, semi-liquid state and digestion proper can now commence. The intestines (see Fig. 2) comprise the duodenum, the small intestine and the large intestine. The small intestine is 'small' because of its smaller diameter; it is actually much longer than the large intestine. Where the two join there are two blind ducts known as the caeca or caecal tubes.

The duodenum

The duodenum has the form of a tightly U-shaped loop which encloses the pancreas. The latter secretes further digestive juices and passes them via the two or three pancreatic ducts into the duodenum for mixing with the material coming down from the gizzard. These slightly alkaline juices contain three ferments or enzymes which, between them, are concerned with further digestion of the three main components of the diet: proteins, starch and fats. The protein enzyme is trypsin; that for starch is diastase or amylopsin; and that for the fats lipase or steapsin.

These pancreatic secretions are of vital importance. If their supply is cut off or impaired the chicken is likely to die within a few days. One such failure came to light when untreated soya beans were used as the major protein source in poultry diets.

Roasted or treated soya beans are an especially useful protein, but uncooked beans contain a trypsin inhibitor which effectively blocks the work of the pancreas in this particular respect.

The small intestines

The small intestine is lined throughout with small finger-like projections known as villi and they give the lining the feel of velvet to the touch. The villi are made up of blood vessels and vessels of the lymphatic system known as lacteals and are covered by very delicate absorptive tissue (epithelial tissue) through which chemical components of the food start to be absorbed into the body.

Food is moved down the length of the intestine by the wave-like contractions and dilation of the circular muscles of its external wall. This characteristic and highly important muscular action is known as peristalsis and, as well as providing directional movement, it also continues to mix the food as it goes along, bringing all particles in contact with the absorptive tissue, so that all the nutrients to be absorbed are taken up during this part of their journey.

The large intestine

The large intestine commences at the junction of the caecal tubes. Its role is to start the accumulation of the indigestible residues coming from the small intestine and its entrance is guarded by sphincter muscles reminiscent of those controlling the crop. Some excretion of soluble salts (e.g. calcium and phosphorus) takes place in the large intestine and these are mixed with the indigestible fibrous residues of the diet. This mixture is now moved towards the cloaca and, with the addition of some water and mucus, forms the droppings.

The caeca

The two caecal tubes of the adult fowl may be as much as 15 cm (6 in) long. Their role in the system is not well known. They have been thought to be responsible for the digestion of fibre – as in other animal digestive systems – but this is no longer important in a chicken normally given little fibre in its diet. There may also be limited synthesis of some of the B vitamins, and the caeca appear to be concerned with the absorption of water. Certainly they become readily filled with liquid and semi-solid material and their content of these has to be expelled three times a day.

Coccidiosis

More important is the fact that the caeca seem to provide an ideal environment for the proliferation of one of the species (*Eimeria tenella*) of coccidia which give rise to the highly lethal disease known as coccidiosis. Other species of *Eimeria* may invade other parts of the intestines and cause serious damage. They do not usually give rise to such massive haemorrhages as can quickly be seen in the caeca as a result of an attack of *E. tenella.*

Once coccidia establish themselves, they not only parasitize the host tissue in such a way as to cause death, but they also have a rapid generation turnover as they shed their eggs in the droppings of the host, and these are liable to be picked up in infective form by other birds picking on the floor or in the litter for food. Fortunately coccidiosis can be controlled effectively by medication but one way of minimizing infection is to deny birds access to the droppings of their fellows by keeping them on wire or slatted floors through which the faeces will have fallen out of reach.

The cloaca

The cloaca is the large chamber at the end of the digestive tract situated immediately before the vent or anus. Several body organs have their outlet in the cloaca. Thus, urine comes from the kidneys via the ureters; the reproductive tract opens into it. Additionally, the pear-shaped 'Bursa of Fabricius' is found in the upper wall of the cloaca in young fowl up to the age of nine months. The full function of this structure has yet to be determined, but it is composed of lymphoid tissue and, in addition to the secretion of lymph, it is concerned with the production of antibodies.

POST-ABSORPTIVE DIGESTION

Outside the alimentary canal lie a number of other organs which are intimately concerned with the processes of digestion and food use. They act as storehouse, purification plants and regulators, and sometimes have direct feed-back connections to the digestive tract itself.

The liver

The liver (see Fig. 1) is involved with several metabolic processes; as a regulator of the concentration of nutrients in the bloodstream, in the detoxification of various substances that may be absorbed into the body and, finally, in the production of an agent of primary digestion – the bile.

To take the last first and relate it to the processes already described, bile is a yellowish-green fluid produced by the liver and stored in the gall bladder – which lies between the two lobes of the liver – prior to its entry into the digestive tract through two bile ducts at the end of the duodenum. The bile contains salts which play a special part in the digestion of fat, turning it into an emulsion in which the lipase of the pancreatic juices can more readily perform its part. The bile also has a mild purgative action, aiding peristalsis of the gut and preventing putrefaction of the gut contents. For its other functions, the liver depends on its supply of blood from the hepatic artery for its oxygenated blood and on the portal vein which brings in the nutrients originally secreted by the villi of the intestines.

Consequent upon this, the liver is concerned both in the synthesis of protein and in the breakdown of excess and non-essential amino acids which it eliminates as urea excreted through the kidneys. It also acts as a store for excess sugar; surplus energy is stored by the body either as glycogen in the liver or muscles, or as fat at many locations but often abdominally.

The liver stores vitamin A on a cumulative basis and the amount present is related directly to the dietary level of vitamin A. Vitamin B_{12} is also stored in the liver.

The kidneys

The kidneys resemble a 'mulberry-like' mass of soft folded tissue occupying the bony hollows on the underside on the roof of the body cavity. Their function is mainly one of excretion, accompanied by filtering out the waste products from its mostly venous blood supply. These products are surplus moisture excreted as

Table 3 Summary of common abnormalities of the digestive system

ORGAN	SYMPTOMS	CAUSE
Beak	Rubbery	Vitamin D deficiency
Mouth	Pasted with hard food under tongue	Food too finely ground
	Tongue inflamed	Niacin deficiency
Throat	Whitish spots	Vitamin A deficiency
	Whitish nodules	Fowl pox
Crop	Impacted	Depraved appetite
	Perforated	Consumption of sharp fibrous material
	Sour	Fermentation of mouldy food by *Candida albicans*
	Pendulous	Weak musculature
	Unthriftiness	Worms
Glandular stomach	Impaction	Accumulation of fibrous material possibly in the absence of insoluble grit
Gizzard	Erosion	Deficiency of vitamin K may be involved
	Abnormal softening or hardening	Use of too finely ground food possibly in absence of insoluble grit
	Unthriftiness	Gizzard worms in ducks and geese
Pancreas	Inhibition of trypsin production	Improperly 'cooked' soya-bean meal
Liver	Fatty degeneration	Shortage of biotin
	Necrotic areas	Blackhead
Caeca	Haemorrhages	Coccidiosis *E. tenella*
Intestines	Haemorrhage, exudate and white spots	Coccidiosis *E. necatrix*
	Red and white spots	Coccidiosis *E. maxima*
	Whitish coating	Coccidiosis *E. acerviculina*

urine and also solid materials such as urates – white crystals of uric acid often seen on the voided droppings as a whitish cap or crust.

THE BLOOD AND BODY FLUIDS – TRANSPORT AND CONTROL

The blood

The blood plays a most important part in metabolism. It collects the nutrients from the villi of the intestines and from the lymphatic system and transports them to tissue sites throughout the body. It also transports hormones (specific 'production controllers') and removes waste products from the production sites. In addition it helps to regulate body temperature and the moisture content of the body tissues.

Blood represents a large fraction of the total weight of a bird; thus a bird weighing 1 kg might have a blood mass equal to 11·6 per cent of her total weight, whereas for one of 2 kg the value might amount to 7·3 per cent.

Blood is made up of semi-solid serum carried in liquid plasma. The main constituents of the serum are the red corpuscles (erythrocytes) and the white corpuscles (leucocytes). Red blood cells occur at the rate of approximately 3 million per cubic millimetre and transport haemoglobin to all parts of the body. They play an important part in general health and a reserve is stored in the spleen for release in an emergency. Leucocytes are much fewer, about one to every 200 erythocytes. When they are more prevalent this is usually an indication of disease of which leukaemia, pullorum and typhoid are noteworthy examples.

Chemical analysis of the blood will give values for its content of glucose and lipids, proteins, non-protein nitrogen, urea nitrogen, uric acid, creatine and creatinine, calcium, phosphorus, chlorides, potassium and alkali reserve. Values will vary with the site of the sample and the physiologically dominating processes at any time.

Water

The amount of body water in the fowl is as high as 70 per cent in the new-born chick and 52 to 55 per cent in the adult. Poultry of all ages and species require a constantly available and clean supply of water from as soon after hatching as is possible. In the early stages it is possible to replace this with separated milk or by a mixture of milk and water if this should be economic. The water should always be readily available and it is an advantage during very hot weather to provide a cool supply whereas during very cold weather it may be necessary to warm the water.

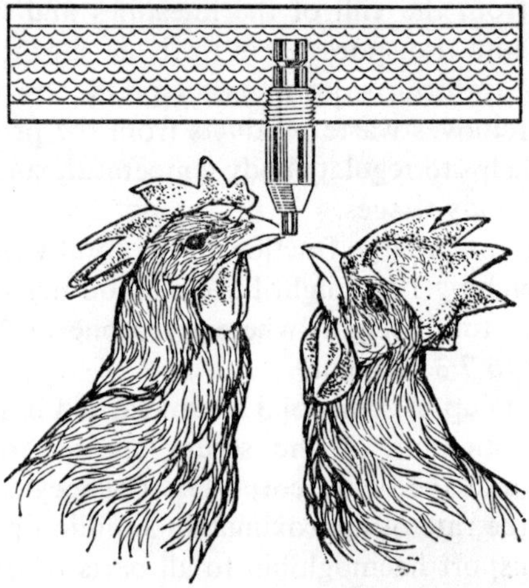

Fig. 4 Nipple drinking trough. Nipple drinkers may be used for poultry of all ages. The supply of water is controlled by a valve. The materials used in construction should be non-corrosive and rust-proof

Whenever possible an automatic supply should be installed, but whether the supply is automatic or not, it is usually necessary to clean the water troughs at least two or three times a week. If these are neglected they quickly become slimy and the birds will avoid drinking. Failure to encourage sufficient water intake will be reflected in poor growth and in indifferent egg production. A laying hen will normally drink 200 cc of water for every 100 g of food she eats. Only in very hot weather may she over-consume water (by a factor of 2·5 times) in order to reduce any threatened rise in her body temperature.

In those cases where birds are kept extensively, water troughs should be moved daily in order to prevent the land immediately around the trough becoming wet and sour. The design of the troughs should be such as to prevent the fowls from perching either on top or on the sides, otherwise they quickly become fouled by droppings.

Another consideration with birds kept on range is that they may find other sources of water. As natural sources of water cannot really be controlled, it is generally wise to fence them off to prevent their becoming a source of disease.

Water is intimately concerned with all body processes. It is the main constituent of blood, lymph, mucus and the digestive juices. It helps to control the body temperature and also plays a vital role in digestion and excretion.

The apparent requirements of birds will to some extent vary according to the composition of the food. Thus, where potatoes and green feed form a large proportion of the diet the birds may appear to drink less. Potatoes contain approximately 75 per cent moisture and green food up to 85 per cent. This point is of particular importance when certain medicinal products are used in the drinking water, for these other sources of water may reduce the total intake of the medicine to below the desired level.

CHAPTER TWO

Fuel supplies – the concept of energy

Introduction

The nutritional processes of the fowl have been likened in Chapter One to the organization and lay-out of a factory. For the factory to operate, fuel must be provided both for driving the individual machines (organs of digestion) and for providing the background environment (e.g. heating) in which the machines will operate. Quite apart, therefore, from the biological function of any part of the system, it must first have energy to make it work and to ensure its continuous running. It will at once be apparent that these fuels are very important nutrients indeed and their supply and the way in which they are used are issues of overriding concern to the nutritionist.

Sources of energy

Fowls obtain their main supplies of fuel to maintain body heat and to meet their other demands for energy from two groups of chemical substances, i.e. from those known collectively as 'carbohydrates' and from 'fats and oils'.

Carbohydrates

Carbohydrates are so named because they are made up of the elements carbon, hydrogen and oxygen, with the hydrogen and oxygen combined in the same proportion as in water. When they are oxidized, they 'burn', releasing heat, carbon dioxide and water according to the general formula:

$$CH_2 + O_2 \rightarrow CO_2 + H_2O + \text{heat (energy)}$$

However, not all carbohydrate-rich foods are equally productive as sources of energy for the chicken, for many quite indigestible materials (such as cellulose) fall into this chemical category. The digestive tract of the fowl is designed only to extract the energy of the more simple carbohydrate molecules. It is like saying that the digestive processes in the chicken are similar to those of a racing-car engine needing high octane fuel. The chicken has no special mechanism like the rumen of the cow with which to change low-grade into high-grade foods. It is essential therefore to distinguish between the less digestible and the more digestible energy-rich foods.

Many materials are included under the term 'carbohydrate'. These commence with simple sugars (which characteristically contain five or six carbon atoms in each molecule), and range through a great number of more complex substances known by the name of 'polysaccharides' – combinations of sugar molecules. Substances included under this heading are starches, pentosans, celluloses and hemi-celluloses. The fowl can really only make use of dextrines, sugars and starches. The remaining carbohydrates appear to be of little value to the chicken except perhaps for some pentosans. By contrast the cereal grains (e.g. maize, wheat and barley) with their content of starch, may make up 70 per cent or more of the total ration of poultry.

The carbohydrate portion of the feedstuffs is expressed in first-level chemical analysis as nitrogen-free extract (NFE). That is to say that it is the portion of the digestible matter of the feedstuff left after fats are dissolved out, nitrogen is extracted and inert ash and fibre have been allowed for. In the case of cereal foods, this figure gives a reasonable indication of the amount of digestible carbohydrate contained, representing primarily the starch content of the flour. In the case of fibrous foods, e.g. grass meal, the nitrogen-free extract figure does not represent the digestible carbohydrate because unfortunately these feedstuffs contain indigestible polysaccharides which are included in the calculation. The celluloses are not included in the nitrogen-free extract, but are separated under the heading 'crude fibre'.

It will at once become obvious that care is necessary when considering an analytical report of a ration. It is very easy to draw false conclusions unless the nitrogen-free extract figure is considered along with its restricting limitations. The figure is arrived at by calculation and is not a direct analysis. As noted above, the figure is in fact obtained by difference after quite other determinations have been made.

FATS AND OILS

Fats and oils are the other group of nutrients which help to supply the fowl with energy. They are always of vegetable or animal origin. There should be no confusion whatsoever between these nutrients and the mineral oils like paraffin and petrol, which can be most injurious. Fats are also compounds of carbon, hydrogen and oxygen, but are much more concentrated than carbohydrates in that they contain far less oxygen. As a result they can use more oxygen to 'burn' more fiercely. Compared with carbohydrate, weight for weight, one part of fat will yield 2·25 times more energy.

Fats and oils are the esters (organic salts) of glycerol and fatty acids; the fatty acids may be classified as being saturated or unsaturated in respect of their chemical structure. Unsaturated fatty acids have a lower melting point than saturated fatty acids. The former, therefore, are often oily liquids at normal temperatures; the latter frequently solid. Fats of animal origin are made up largely of saturated fatty acids whereas those of plant origin are unsaturated.

The saturated fatty acids are far more stable than the unsaturated fatty acids which may rapidly develop oxidative rancidity; i.e. degradation involving oxidation by atmospheric oxygen. Rancid fat may, when mixed with other food ingredients, act upon the fat-soluble vitamins to such an extent that their value is lost. Vitamin A, D and E may be depleted in this manner.

Thus when animal fats and vegetable oils are included in poultry foods, the keeping qualities of the product will depend upon the type of fatty acids that make it up. The greater the proportion of unsaturated fatty acids the greater will be the speed at which oxidation can take place. The rate of oxidation can be delayed by including an antioxidant (see Chapter Six) in the product as soon after manufacture as possible.

However, a small percentage of fat in the diet of all classes of poultry is desirable as some fats act as carriers of certain vitamins known as 'fat-soluble' vitamins (see Chapter Four). Additionally, two unsaturated fatty acids are considered as essential nutrients for poultry. These are arachidonic and linoleic acids and neither can be synthesized by the bird. At one time these two compounds together with linolenic acid were called vitamin F.

Too much fat in the formulation of the diet may give problems with the physical form of the ration and the way in which it can be handled. In general, fats and oils are valuable sources of supplementary energy and are particularly useful for increasing the energy density (see page 47) of rations, the intake of which may be suboptimal for one reason or another – for example, in the case of broilers.

Chemical analysis can show the percentage of fats and oils in the diet under the individual heading of 'oil' or 'ether extract'. The percentage is calculated by dissolving a portion of the feedstuff in petroleum spirit and measuring the result. The figure obtained is not absolutely correct, in that it does contain other substances which are dissolved in the spirit, e.g. free fatty acids, fat-soluble vitamins and pigments. The quantity of these fat-soluble substances is not usually very great, but they are often of outstanding importance. There is a group of substances, sterols, that are dissolved by the petroleum spirit and can be included under the term 'oil'. In the case of vegetable matter, the sterols are known as ergosterols; in the case of animal matter, cholesterols. The sterols are very vital substances, for they are concerned in the direct production of vitamin D (see Chapter Four).

Yet another group of substances may be embraced by the term 'oil' fraction. These are the colouring pigments, especially carotenoids. The pro-vitamin A, known as carotene, is included in this group. This substance is called pro-vitamin A because from it animals have the ability to build up vitamin A (see Chapter Four).

In the case of cereal foods, the term 'ether extract' or 'oil' may indicate practically true oil. However, protein-rich feedstuffs and grass meal are a different matter, and the fat-like substances included in the calculated 'oil' content of an ingredient may well reach as much as 50 per cent.

FIBRE

The term 'crude fibre' embraces the insoluble portion of the feedstuff, e.g. cellulose, which is almost valueless nutritionally. Too much fibre in the diet should therefore be avoided for this renders the food indigestible and low in available energy. The optimum level of fibre lies between 2 and 3 per cent and the normal maximum is 5 per cent in starter, grower and layer rations. The higher the level of the fibre the bulkier the ration becomes and therefore the supply to the bird of the digestible portion of the diet is restricted. Conversely the lower the fibre content the more efficient will be the conversion of feed into eggs or meat.

The crude-fibre content of the food is estimated by treating it first with alkali and then dilute acid in accordance with a method devised in the middle of the nineteenth century. The analysis was made originally for cattle foods and can therefore have only a limited application for poultry. The digestive juices of the chicken do not contain enzymes capable of breaking down fibre to any extent, although some breakdown may perhaps occur in the two blind caecal tubes. Fibre is, however, thought to play a part in helping peristaltic movement in the digestive tract and thus assists in maintaining muscle tone.

Rations containing from 10 to 20 per cent fibre may, on occasions be suggested for special purposes, e.g. holding rations for potential breeding stock, when it is desired to limit the intake of nutrients and reduce growth rate (see Chapters Eight and Nine).

The measurement of energy values

Many figures have been used to denote the energy content of a ration. They vary chiefly in that they may describe the ration in absolute terms or in its comparative ability to supply energy to the chicken for particular purposes. It must be emphasized, therefore, that the figures giving energy levels are only a guide, for the ultimate way to calculate the feeding value of an ingredient or food mixture is by means of feeding trials with the class of stock for which the food is to be used. Rations, however, must be formulated before they are fed to chickens and for this reason some satisfactory way of designating the potential energy-producing content of an ingredient must be utilized. Some of these are reviewed below.

NITROGEN-FREE EXTRACT (NFE)

In the past, nutrition chemists have referred to the carbohydrate portion of a food as the nitrogen-free extract. This is an indirect calculation rather than a precise determination. In the case of cereals the NFE gives a reasonable indication of the amount of digestible carbohydrate, but in the case of the fibrous foods, e.g. alfalfa meal, it does not. Fibrous materials contain indigestible carbohydrates, known as polysaccharides, and these are included in the NFE. The cellulose fibres themselves are not included in the NFE calculation but are isolated under the term 'crude fibre'. For these reasons the NFE figure is not of great value to the poultry nutritionist. More direct estimations are required.

GROSS ENERGY

The gross or potential energy of a feedstuff is estimated by burning the sample in a bomb-calorimeter and measuring the heat produced. This is an absolute chemical determination and it is hardly to be supposed that a bird's digestive system will be as effective a mechanism for liberating the whole of this energy supply. Gross energy, however, is the logical starting-point.

DIGESTIBLE ENERGY

Once food is consumed, a portion is in fact digested and the remainder (mostly indigestible but also some digestible material) is voided in the faeces. The term 'digestible energy' (DE), therefore, refers to the 'gross energy' (GE) less the energy of the faeces and the waste gases.

$$\text{DE} = \text{GE} - \text{GE of faeces} - \text{GE of waste gases}$$

METABOLIZABLE ENERGY

Even when digested, some of the energy is still lost through inefficiencies in the process and is voided in the urinary system. In the case of fowls, the urinary waste products are voided as urates and are seen as a white cap on the faeces.

Metabolizable energy (ME), therefore, denotes the digestible energy less the energy lost by way of the urinary system. In chickens, compared with other animals, it is comparatively easier to estimate the metabolizable energy rather than the digestible energy, because the waste products of the intestines and the urates are excreted together. Thus:

$$\text{ME} = \text{GE} - \text{GE of faeces and urates} - \text{GE of waste gases}$$

This value is the one most frequently used to describe the energetic value of ingredients and complete mixes for chickens. In consequence, tables showing the metabolizable-energy value of most feedstuffs have been compiled and are available (see Table 50) to help compute rations to meet the established energy requirements of all types of domestic poultry. Such values are a necessary and useful first indication of energy potential but they should always be checked through, if at all possible, by direct feeding trial determinations.

The energy content is expressed in terms of some agreed unit of energy measurement per unit weight of food. This has most frequently been done in terms of calories per pound (or now, in metric terms per kilogram).

A calorie is, of course, the amount of heat required to raise the temperature of 1 cc (1 g) of water through 1 degree at 15° C. Since the normal unit weight of rations is either kilograms (or pounds) and the body weight of a chicken is likely to be more than a kilogram, some larger unit than a calorie is required unless the figures are to be in millions. Accordingly, the term kilocalorie has frequently been used, giving, for example, for maize an average ME value of 3,300 kcal per kg. More recently it has become usual to write this in megacalories, i.e. 3·3 mcal/kg.

Some workers have thought it desirable to describe the rate of heat production of a food, since not all ingested food is converted at once. The unit used to describe this is the joule which is the output in one second of an energy flow of one watt. For most practical purposes, however, we can accept a simple conversion of joules to calories and vice versa from the equation 1 kcal = 4·18 k J.

NETT ENERGY

Whilst metabolizable energy represents the energy available to the bird, not all is used for production, which is two stages away. Nett energy represents the metabolizable energy less the energy given off as heat by the body in setting the metabolic process in motion. This is sometimes referred to as the 'specific dynamic effect', 'energy of utilization' or 'heat increment'. The nett energy figure represents the energy available for maintenance activities, growth or egg production.

In this connection it is worth noting that some feedstuffs which are naturally rich in oil are undervalued by their theoretical ME value. For many metabolic purposes, the ME in the oil fraction of, say, maize is more efficiently used than the starch ME in the grain. The latter has to be broken down through various intermediate processes and then built up again to form the triglyceride molecules used to construct fatty tissues. This process consumes much energy, showing a high 'heat increment'. By contrast the fatty acids in the oil, particularly those with an extended carbon chain length, are readily used to form the same triglycerides directly and less of the original energy supply is wasted. This information can be utilized by the poultryman, especially the turkey producer.

PRODUCTIVE ENERGY

Productive energy may be described as the energy actually used for the end point productive purposes, e.g. egg production or growth. The figure represents the nett energy, less the energy used for maintenance.

At first it might appear that the productive energy is the most satisfactory method of measuring energy values. Unfortunately this is not so, since it is really measuring the efficiency of a particular production system, in which, for example, an undue amount of heat might be used in maintaining body temperature in a badly insulated poultry house. This would not be a characteristic of the ration itself and it is generally more convenient to use the ME value which does describe the ration and then to assess the efficiency of the latter's use as a function of the productive system.

If the production system is genuinely standardized then it may be useful to compute the productive-energy value of a ration in the way that many research workers, notably G. S. Fraps in the United States, have done.

Energy requirement

The example mentioned in the section on productive energy makes the point that birds may 'require' energy for seemingly unproductive processes. In this instance a particular need was seen as a function of the environment in which the bird lives and which we might, not too fancifully, talk about as the energy costs for the bird to run its own central heating system. And there are other energy needs too: that of getting the fuel to the burners, that of overcoming the potential inertia of the system and that of covering its mechanical operation. All have to be satisfied to sustain the bird in a state of basal metabolism (or to get the factory ready for production).

Beyond these, energy is required for the undemonstrative functions of daily living, breathing and movement and then, finally,

Table 4 The energy and fatty-acid value of whole egg minus the shell (*Poultry and Egg National Board, 1964, USA*)

Nutrients per 100 g	Water	75·5 g
	Calories	163 g
	Lipids	11 g
	Fatty acids:	
	saturated	3·6 g
	mono-unsaturated	6·2 g
	poly-unsaturated	1·4 g
	Cholesterol	0·8 g
	Phospholipids	3·3 g
	Carbohydrates	0·7 g

for the productive processes of changing raw materials (proteins) into liveweight gain and into egg formation. Fuel (energy) is needed, therefore, for many purposes and the efficiency of its use will clearly depend on many factors. By the same token, it will be clear that it is not possible to make a single estimate of the bird's energy requirement for, say, egg production (even given the information in Table 4) until many aspects of the production system have been detailed. This fact leads to some important preliminary decisions, as may be seen below.

Factors affecting energy requirement

ENVIRONMENTAL TEMPERATURE

Chickens are 'homeothermic', which means that they attempt to regulate their metabolism to maintain a stable body temperature at about 41·7° C. If the temperature in the poultry house is substantially below this (and the chances are that house temperature will be in the range of 12° C to 22° C) then an important fraction of the bird's food intake will be devoted to maintaining the differential between her body temperature and that of the air which surrounds her. Conversely if the house temperature rises, either because it is warm outside the house or because the insulation properties of the structure are designed to accumulate the heat production of the birds, then the food required for body-heat maintenance will fall, with the usual result that food intake by the bird will also fall.

This is an important issue in cool climates and especially when feedstuffs may be more than usually expensive; there is a strong case for conserving as much of the natural heat production of the birds as possible. This may be done by better poultry-house construction, especially in providing improved insulation properties, and also by better control of ventilation, reducing the entry of cold outside air to the level that will ensure cleansing of the air of carbon dioxide and ammonia whilst not at the same time removing too much heat from the house.

In hot climates the problem is obviously different. In order to adapt to the increased environmental temperature the bird's food intake falls, perhaps to a dangerous degree when the amount of the ration eaten no longer contains sufficient nutrients to sustain normal production, perhaps even to sustain life itself. Death following heat prostration is far from uncommon in poultry flocks in tropical and sub-tropical areas.

Clearly then, energy requirement is affected by environmental temperature and, as a result of the relationship, energy intake is likely to vary to a higher degree than energy requirement (especially if the latter is related to a given end point of production). If too little intake of energy occurs, too few nutrients may be provided for, say, yolk development. If too much energy is ingested, some may be recovered in increased egg size but most will be wasted (economically speaking) in heat production or in storage as fat in the body tissues.

Disease

One characteristic of a wide range of diseases is the fever which they may produce and the loss of appetite which results. The consequences of reduced food intake have already been made clear. An opposite effect may occur if birds become heavily infested with parasitic worms. In this case requirement will seem to be high, for birds will have a comparatively high intake of food but yet be far from normally productive.

APPETITE

Estimated requirements for different classes of birds for different production functions are discussed in later chapters (see Chapter 8) but it will now be clear that nutrients to supply these requirements will have to be contained within the limits of the food intake of the bird in various sets of circumstances. This is the critical point.

Appetite varies with health, size and level of production of a bird. A modern egg-laying hybrid will eat to an average intake of approximately 110 g per day. Within a normal range of temperature in a poultry house in temperate latitudes, that intake may vary by plus or minus 15 to 20 g without serious consequences. If we accept another average figure for the same bird, that representing her energy requirement, then this would have a value of about 300 to 330 kcal per day. Clearly then, her diet must carry nearly 3 kcal per g of food eaten.

This value is readily achieved if a good-quality mixture of cereals provides the bulk of the ration, and is a first guide to what is meant by the 'energy density' of the ration – the proportion of energy to unit weight of food. Rations may be more or less dense according to the way in which, on the one hand, fats are added to the diet or, on the other, cereals (or other ingredients) of lower than normal energy content have to form the bulk of the diet.

If appetite is depressed, for whatever reason, by a large amount – to, say, 70 g per day – and the ration is formulated to give average energy density, then the bird can take in only about two-thirds of her daily requirement of nutrients and performance must suffer. To sustain it one would, theoretically, need either to reduce the bird's requirement (e.g. by warming the environment) or to increase the density of the ration so that reduced physical intake would still give normal ingestion of nutrients.

If rations of lower than normal density are fed, the bird may have the capacity to eat more each day in order to match intake to requirement. In general terms it would be fair to say that appetite is regulated (by mechanisms not completely understood but monitored by the hypothalamus in the brain) to achieve just such a balance. Small departures from balance may also be buffered by storing energy in the body as fat and then remobilizing the stored energy in times of deficit.

One situation where external conditions may seriously affect this balance by reducing appetite is that of high temperatures (those in excess of, say, 27° C). In this case a ration, whilst perhaps not needing to be more dense in energy, would certainly need to be formulated with higher proportions of protein, vitamins and minerals which would otherwise be certainly eaten at less than optimum rates.

ENERGY–PROTEIN RATIOS

Except for the very last point above, we have let the idea of the energy density of the ration stand for the broader concept of its nutrient density. In effect we have said that we can adjust the grade of the fuel we bring to the machine floor to compensate for variations in fuel uptake. Again in broad terms this is a useful generalization and we can expect that the energy level of the ration will be closely related to that of the other components.

This relationship is known as the energy–protein balance of the ration and is usually denoted by the number of energy units per cent of crude protein in the mixture. What this really tells us is the amount of energy in the rations (over nett energy and energy needed for maintenance) that is required to metabolize the protein supplied to body tissues. At an elementary level of interpretation it does this quite usefully and it has been possible in the past to demonstrate marked differences in growth performance by the bird, depending on differences in energy–protein balance above or below what is thought to be the optimum. On this basis energy–protein ratios for different classes of stock are quoted in later chapters.

However, as will be seen in Chapter Three, the crude protein percentage is a somewhat insensitive description of the protein content of the ration, and the idea of the energy–protein ratio as a characteristic of the mixture is only valid when more precise specifications of protein requirement and/or response to protein have been made. The energetic cost of metabolizing protein is high and it is almost certainly necessary to estimate the amount of protein production in the body that can be economically expected per unit of time, before suitable ratio figures can be calculated. The possibility of quite large departures from normal balances to meet the requirements of particular sets of circumstances has already been noted (e.g. in hot climates).

The effect of the physical characteristics of the ration

We have talked about the energetic regulation of intake by the bird, but another important factor governing her use of foods is their physical form. This is an important practical point and is dealt with in detail on pages 197–207. In general, finely divided and dusty foods are not readily eaten and can lead to a marked drop in intake. Very fibrous foods (like a ration based on ground oats) are too fluffy and carry a low proportion of nutrients per unit volume rather than per unit weight. Both of these problems can be eased if the ration is presented not in raw ground form but after it has been pelleted. With the use of pellets, intake of food and also intake of energy (despite natural regulation mechanisms) can be increased by 8 to 10 per cent. This is one reason why crumbs and pellets are very popular in feeding broiler chicken.

CHAPTER THREE

Raw materials for manufacture – the role of proteins

Introduction

The building blocks of a growing or production metabolism are the nitrogen-bearing chemicals known as proteins. These are essential for the proper growth and general well-being of all classes of fowl, for proteins are a major constituent of the nucleus and the protoplasm of all body cells.

In addition to nitrogen, proteins are made up of carbon, hydrogen, oxygen, and, in some cases, phosphorus. They are usually very complex molecules made up of more simply recognizable units in varying combinations. These units are called amino acids and each has its own characteristic chemical composition.

The proteins in the body of the chicken or in the eggs she lays are made from the amino acids assembled in the gut in the process of digestion and then modified in other organs of the body. It is obvious that these ingredients or units must come from the foods that are offered and these plant (and sometimes animal) materials have their own similarly characteristic combinations of amino acids. Protein nutrition, therefore, consists in identifying the amino-acid requirements of chicken and then in supplying these directly (essential amino acids – see page 53) or in supplying feedstuffs which will first be broken down into simpler nitrogenous compounds from which, in turn, the animal's amino-acid requirement can be synthesized. Amino acids that can be synthezised in this way are called non-essential amino acids (i.e. it is not essential that they be supplied directly in the diet).

The estimation of the protein value of feedstuffs

Most materials used as ingredients for poultry feeds contain some protein. Thus grains may well contain 8 per cent of crude protein (see page 280) in their dry matter. If cereals were the sole constituents of the diet a daily intake of 110 g of food would yield 7 to 8 g of protein which would represent about half the daily requirement of the adult chicken. It is, therefore, always necessary to include some ingredients which have a much higher concentration of protein. As might be imagined, these come from areas of rapid growth or productive activity in either plant or animal sources. Thus grass or lucerne leaf and seeds (especially of legumes) when the oil content has been extracted are good sources of vegetable protein, whilst dried fish or meat offals (fish meal, meat meal, blood meal) supply animal protein. These products might have a protein content, in the dry matter, in the range 20 to 70 per cent.

PERCENTAGE CRUDE PROTEIN

It is necessary, then, to have a measurement which will distinguish quickly between materials of high protein content and those which are not. This measure is the percentage content of crude protein or 'albuminoids'. The figure is obtained by calculation based on the fact that proteins generally have a nitrogen content in the region of 16 per cent. The nitrogen content of the feedstuff is estimated chemically and this figure is multiplied by 6·25 (100 divided by 16 = 6·25). The assumption is that all the nitrogen determined in the test is present in the feedstuff as protein. This is by no means necessarily the case. It may, therefore, be necessary to calculate and allow for the 'non-protein' nitrogen present.

The crude-protein content of a feed is, therefore, only a very general guide to its nutritional value for it has no sound nutritional basis. The concept is, however, well understood and has a use in making simple comparisons between feedstuffs. Its limitations are important and must be clearly understood. Thus:

(a) Not all protein is equally digestible and a first need is to find out the digestible crude-protein content.

(b) 'Crude' protein tells us nothing about the constituent amino acids and, therefore, about feeding value. Thus a ration could show on analysis to have a crude-protein percentage of 18 per cent, but lacking an essential amino acid it might be much less productive than a better-balanced mixture containing only 12 to 14 per cent crude protein.

PROTEIN QUALITY

An accurate and rapid method of determining the quality as well as the percentage of protein in a food ingredient is required in order to allow the right decision to be made before the food is used for animal feeding. This is not readily available since a judgement of quality really requires a full knowledge of the amino-acid content.

In practice the majority of animal food ingredients are purchased on the 'crude protein' content; this is mainly due to custom, and to the lack of internationally agreed methods of evaluating protein quality for the purpose for which it is to be used.

Whilst feedstuffs rich in animal protein vary considerably in their amino-acid make-up, and in the availability of certain of the constituent amino acids, it is generally true that these feedstuffs are of a higher biological value, i.e. have a superior amino-acid balance than the vegetable protein-rich foods, and this has led to the widespread practice of including a minimum of 3 per cent of animal protein in poultry food mixtures.

In the case of some feedstuffs of vegetable origin, the amount of non-protein nitrogen may represent almost half of the value calculated as crude protein. In the case of animal protein-rich foods and cereals, the total estimated protein content consists of almost 100 per cent pure protein.

However, the superiority of animal protein-rich feedstuffs over the vegetable protein-rich foods is not entirely a question of amino-acid content. The animal protein sources are often carriers of some of the essential vitamins and minerals. Fish meal, for example, contains vitamin B_{12} and also higher levels of certain minerals which are not found in vegetable protein foods. For this reason, when a high percentage of a vegetable protein food, e.g. soya-bean meal, is included in the diet, it is necessary to check the vitamin and mineral content of the diet extremely carefully.

AMINO-ACID COMPOSITION

It is necessary, therefore, to know the composition of any protein-rich feedstuff in terms of its constituent amino acids.

At least twenty amino acids have been recognized, of which ten may be considered as being essential for poultry. The essential amino acids, required for growth and other purposes, which cannot be synthesized by the bird, are:

Arginine	Methionine
Histidine	Phenylalanine
Isoleucine	Threonine
Leucine	Tryptophan
Lysine	Valine

Of the other acids, tyrosine may be formed in the tissues from phenylalanine and cystine similarly from methionine. It is common, therefore, to consider meeting a combined requirement for phenylalanine + tyrosine and one for methionine + cystine. Deficiencies in supply of the second acid in each pair can usually be met by an extra supply of the first. By contrast a surfeit of the second acid cannot make good a deficiency of the first.

Glycine (usually considered with serine which probably represents part of the pathway for the synthesis of the former) and proline are synthesized in the body, but synthesis may be inadequate during periods of very high growth rate when these substances may have to be regarded, temporarily, as 'essential'. Alanine, aspartic acid and glutamic acid are synthesized in the body in adequate quantities and are, therefore, non-essential.

Cystine and methinonine contain the element sulphur in their make-up and are sometimes referred to as the sulphur-bearing amino acids.

The amino-acid content of the different protein sources varies considerably but values are now available for the most likely ingredients. In the event of the essential amino acids being present in insufficient quantities they will act as limiting factors to the biological value of the protein as a whole, and this may be revealed in the form of poor growth, poor egg production or poor conversion.

Balancing the amino-acid inputs to ensure that the essential ones are present at the correct levels is one of the most important factors in formulating food mixtures. Unfortunately, the levels of the amino acids occurring in food ingredients do not match the bird's requirements and some may be in excess and others deficient.

It follows that the balance between component amino acids has to be struck according to the bird's requirements at different times.

CHEMICAL EVALUATION OF PROTEINS

Chemical methods of determining the amino-acid value of feed-stuffs are comparatively quick, and automatic amino-acid analysers, built to analyse many samples at a time, are widely used.

Using an automatic analyser, it is possible to determine the amounts of most of the range of amino acids in a feed, within three to twelve hours, according to the complexity of the mixture. If one or two are 'lost' because of the solvents used, separate determinations can be made for these. All the automatic amino-acid analysers work on the principle of the chromatograph, thus giving a visual representation from which the amino-acid values are determined against a pre-calibrated scale.

Chemical methods of assaying amino acids, with the exception of lysine, do not give any indication of the actual amount of the amino acid that is available to the bird. Lysine is particularly prone to become 'bound' or unavailable, but fortunately a chemical method of assaying the lysine value of feed in which the 'bound' lysine is ignored has been developed.

In the case of feedstuffs that have been subjected to considerable heat, e.g. possibly over-heated during manufacture, a good deal of the value of the protein may have been lost. In this case, the total amino-acid value suggested by most chemical means of assay may be greatly in excess of the amount that may be biologically available.

Chemical means of estimating the amino-acid values are not therefore entirely satisfactory on their own. Damage to protein by heat will not be apparent to the naked eye, and may not be revealed by chemical tests.

MICROBIOLOGICAL ASSAY OF AMINO ACIDS

Microbiological methods of assaying the amino-acid values of a food (i.e. by using micro-organisms, usually bacteria) are comparatively quick, taking approximately two days, and give a reasonable indication of the availability of the amino acid.

A number of proteolytic micro-organisms have been found which will digest protein and their subsequent growth will be in proportion to the amino-acid value of the feedstuff under test. In this way, for example, the 'available lysine' or 'available methionine' may be calculated.

In the case of assays made with the micro-organism streptococcus zymogenes, the amino-acid values obtained have been shown to correlate well with feeding tests with chicks, thus pointing the way to the development of a valuable tool in the quest for standard and rapid means of assessing protein quality.

By these means the amino-acid content of a food ingredient can be expressed either as a percentage of the dry matter of the food as a whole, or as a percentage of its protein fraction.

One of the problems in quoting average amino-acid contents is the fact that individual feedstuffs vary considerably in their amino-acid content from sample to sample.

Protein requirement

Balancing the food in respect of the essential amino acids may present some difficulties in view of the number to be considered and the varied sources from which they come.

In particular, poultry derive their amino-acid requirements from cereals as well as from the protein-rich foods. Cereals supply significant quantities of amino acids to the bird in view of their large level of inclusion (60 to 70 per cent) in the total diet.

HIGH-PROTEIN CEREALS

In respect of the fact noted above and also because protein-rich foods are seldom in generous supply and are expensive, considerable scientific endeavour has been made to develop cereals, especially maize and wheat, with higher than normal levels of protein.

Breeders of maize have been successful in developing strains with increased levels of lysine compared with standard maize. The gene responsible for controlling the lysine content of maize is called Opaque-2. Sufficient work has now been conducted to show that the production of maize with a crude-protein content as high as 20 per cent is now in the realms of scientific possibility.

Triticale is another example of the success of the plant breeder in developing a high-protein cereal. Triticale, which has been produced from wheat and rye, has a protein content that may reach over 18 per cent.

THE CONCEPT OF BALANCE OF AMINO ACIDS

Obviously the use of a single source of protein in large quantity in a ration will lead to the dominance of those amino acids which are characteristic of that source or ingredient. There then follows the possibility that some other amino acid, normally incorporated from another source will not be present, or be present in insufficient quantity.

An excess or deficiency of one amino acid can upset the amino-acid balance and the usefulness of the entire feed; for example, an excess of lysine is not only wasteful in itself but may also impair the utilization of arginine, thus increasing the apparent requirement for the latter. Much the same is true of an excess of leucine in respect of isoleucine and valine or of isoleucine in relation to leucine and valine.

It is, therefore, very necessary to formulate rations to achieve an appropriate balance of amino acids, avoiding both deficiencies certainly but also the less obvious surpluses. The question, therefore, is how is this to be done?

THE IDEA OF AMINO-ACID LIMITATION

In recent years much research has been completed which has been designed to demonstrate the response of both broilers and laying chickens to increases and decreases in the content of individual amino acids in the ration. A basal diet is formulated in which the acids, other than the one under trial, are held at what is believed to be an adequate level to meet known minimum requirements for each, and then the content of the experimental ingredient is included at varying levels in order to identify the shape of the 'response' curve to these changes.

In principle, these response curves are similar in form, giving rapid increases in response at levels of input just above the minimum and then a diminishing rate of response as inputs increase still further. Most curves show additional responses to quite high levels of individual amino-acid inclusion (except where there are known antagonisms between one and another) but it will be obvious that as responses decrease, the increased cost of obtaining them may not be economic.

It becomes, therefore, a matter of identifying economic responses and trying to satisfy them. These will vary with different amino acids, and the ration formulator is driven to take note of those amino acids of which a decrease in input (considered, perhaps, because of cost) causes a bigger decrease in performance than a similar unit decrease with the inclusion of others. These are known as 'limiting' amino acids and one tries to identify them in sequence as 'first limiting', 'second limiting' and so on.

Up to the present, methionine and lysine as first limiting and second limiting acids have been well investigated and reliable information is also available for tryptophan and isoleucine. Arginine and valine are thought to have perhaps the next most critical level of inclusion but satisfactory results have still to be produced. It is to be noted that the order of limitation is not related to the absolute content of the individual acids in the egg as set down in Table 5.

The order of amino-acid limitation in a feed will be influenced by the ingredients used. Thus where soya-bean meal supplies the bulk of the protein the first limiting amino acid will usually be methionine. On the other hand, where cottonseed meal provides the protein, lysine is likely to be first limiting.

Table 5 Amino-acid value of whole egg minus the shell (*Poultry and Egg National Board, 1964, USA*)

		g
Nutrients per 100 g	Protein	12·0
	Amino acids:	
	Arginine	0·80
	Aspartic acid	0·70
	Cystine	0·26
	Glutamic acid	1·48
	Glycine	0·44
	Histidine	0·32
	Isoleucine	0·84
	Leucine	1·02
	Lysine	0·82
	Methionine	0·40
	Phenylalanine	0·65
	Serine	0·92
	Threonine	0·66
	Tryptophan	0·23
	Tyrosine	0·55
	Valine	0·98

SYNTHETIC AMINO ACIDS

One of the reasons why much is understood concerning methionine and lysine is because they are available as separately manufactured ingredients at an economical cost, and are now quite frequently included in rations in this form. Other amino acids are available in synthetic form but are still too expensive to consider as ingredients of any save experimental rations.

One approach that these forms of amino acids allows is that the initial balance of the ration can be compounded on the basis of the balance of the less limiting amino acid so that calculated amounts of methionine and lysine can be added at a later stage.

VARIATIONS IN PROTEIN REQUIREMENT

The amounts of protein required by fowls vary considerably according to age and stage of production. The demand for protein during the first six to eight weeks of life, during egg production and the moult are extremely high. Also, protein requirements are closely associated with the energy level of the total diet – the higher the energy level, the greater the percentage of protein that is required.

For example, in the first case, the lysine requirement during the first two weeks of life amounts to 5 per cent of the protein level of the diet, to 4·5 per cent of the protein from two to sixteen weeks, and to 4·0 per cent of the protein after sixteen weeks of age.

The point about the relationship between protein and energy has already been discussed under energy/protein ratio in Chapter Two. Detailed statements and recommendations for different classes of stock are set out in Chapter Eight.

AMINO ACIDS AND VITAMINS

Relationships exist between certain amino acids and vitamins; thus to some extent methionine can spare the vitamin choline and tryptophan can spare niacin.

CHAPTER FOUR

Production regulators: Group One – the vitamins

Introduction

Vitamins are complex organic compounds that play an important role in metabolism. Although they are required by fowls in relatively small amounts, the presence or absence of these vitamins make make all the difference between profit and loss. Many of the vitamins have been isolated and recognized chemically, and a number may now be produced synthetically, so that the part they play is now much better understood.

An adequate supply of vitamins is necessary for the proper functioning of all body processes. Thus good growth, egg production, the hatchability of eggs, and the prevention of leg weakness, all demand a proper vitamin balance and could be said to show a precise vitamin requirement. A diet may be complete in every way except for one essential vitamin, and its absence may cause a corresponding deficiency disease and considerable financial loss.

The vitamin requirements for poultry, as commonly known, are to be found in Tables 16 and 17. However, the requirements of the different species of poultry have not in every case been precisely established. This is an important point for it is possible for a bird to suffer from an excess of certain vitamins as well as, more usually, from a deficiency.

OCCURRENCE AND SUPPLY OF VITAMINS

Vitamins occur in many substances likely to be used as feedstuff ingredients. They may be considered in two main groups: the 'fat-soluble vitamins' and the 'water-soluble vitamins'. The fat-soluble vitamins are A, D, E and K. The water-soluble vitamins are those of the vitamin B complex and vitamin C. The important vitamins are listed in Table 6 and those normally supplied in adequate amounts in poultry foods in Table 6a.

Table 6 Vitamins to be considered in the formulation of poultry foods (*Poultry and Egg National Board, 1964, USA*)

Vitamin	*Content per 100 g whole egg minus the shell*
Vitamin A	1,140 IU
Vitamin B complex:	
Aneurine, thiamine or B1	0·1 mg
Biotin (once B4 or H)	22·5 mcg
Choline	0·53 g
Folacin or folic acid	9·4 mg
Niacin, nicotinic acid (once PP)	0·1 mg
Pantothenic acid (once B3)	2·7 mg
Pyridoxine or B6	0·25 mg
Riboflavin or B2	0·29 mg
B12 or cobalamine	0·28 mcg
Vitamin D	50 IU
Vitamin E	2·0 mg
Vitamin K	trace

Table 6a Vitamins normally supplied in adequate amounts in poultry foods

Vitamin B complex:	
Inositol	33·0 mg
Para-aminobenzoic acid	—
Vitamin C	—

Vitamins may be available in their pure state or in the form of a substance called a pro-vitamin, which is a precursor, that can be converted within the bird's digestive system into a vitamin. The best-known example of this is pro-vitamin A, or carotene, a normal ingredient of green foods, which the fowl uses to build up vitamin A.

VITAMIN INSTABILITY – LOSSES ON STORAGE AND PROCESSING

Vitamins, particularly the fat-soluble vitamins, whether occurring naturally in poultry foods or in special supplements, require satisfactory conditions of storage to delay the process of loss of vitamin value which is liable to occur from a variety of causes.

Vitamin supplements (see Chapter Four, pages 64–65) should

be stored in light-proof surroundings at temperatures above freezing-point, but not exceeding 20° C (68° F). In hot climates the food-additive stores should be air-conditioned and the temperature recorded regularly so that equable conditions can be maintained.

Milling and mixing

Vitamin A and D_3, and to a lesser extent vitamin E, may be rapidly oxidized and destroyed by adverse conditions of heat, light, moisture, contact with rancid fat, and particularly by admixture with certain minerals. Substantial losses occur for these reasons during the process of milling and mixing. To guard against these losses, margins of safety must be allowed in formulation and these usually provide for an inclusion of 30 to 50 per cent more than the estimated requirement for the fat-soluble vitamins.

The potency of vitamin A may be lost in as short a time as two to three weeks if mixed with feeding salt or choline chloride, and it is usually advised that these minerals be added directly to the food mixture rather than included in a vitamin/trace mineral mixture.

The loss of value of vitamin A occurring during the process of mixing may be up to 10 per cent and a further 10 per cent loss may occur during pelleting. If, however, the pelleting temperature exceeds 70° C (160° F) the loss of vitamin value can reach 80 per cent. It is assumed that the wastage of vitamin D_3 in a mechanically mixed ration is similar to that of vitamin A.

Vitamin E is also subject to loss of value once mixed with other ingredients and the wastage during processing is considered to be about half of that of vitamins A and D_3.

The loss of value of vitamin K may be regarded as being about 10 per cent during mixing and 20 per cent in pelleting and from 5 to 25 per cent during a storage period of three months.

Compared with the fat-soluble vitamins the water-soluble vitamins are comparatively stable and, subject to good storage conditions, loss of value is not usually of practical significance. Only the naturally occurring pantothenic acid tends to become oxidized during processing and losses as high as 60 per cent have been reported from wheat that has been ground.

Biotin, choline, folic acid, pyridoxine and riboflavin are very stable products, whether occurring naturally or in the synthetic form, and losses will not usually arise during the process of milling and mixing.

Storage

All the processes to which whole grains are subjected during storage can be expected to lead to a loss of the naturally occurring fat-soluble vitamin E.

Vitamin E present in dry freshly harvested cereals may be in the region of 12 to 16 mg/kg of alpha tocopherol and the value is retained if the grain is properly dried. However, the loss of alpha tocopherol during storage has been estimated to amount to about 10 per cent for each one-month period so that, by the end of the season, the value may have fallen to 4 to 6 mg/kg.

Cereals stored in moist conditions, even if treated with a preservative, rapidly lose their vitamin value. Once ground, the loss of alpha tocopherol can soon reach as much as 90 per cent of that originally present.

Thiamin is sensitive to heat and light. There is, for example, a loss of thiamin during the artificial drying of cereals and subsequent storage which, if the temperature is too high, may reach 25 per cent of the total thiamin originally present.

VITAMIN ASSAY

In order to guard against the losses of these critical nutrients that have been outlined above, it is necessary to gain some reassurance about the vitamin values that are to be used in the making up of a poultry food. This can be attempted by appropriate analysis. In particular, regular checks of vitamin A level and potency should be made. However, vitamin analysis is not a simple matter and specialized equipment and training is necessary for this exercise. The manufacturers and suppliers of ration additives usually maintain an analytical service as well as some kind of testing facility.

Biological testing, using the animals for which the ration is formulated, is the only accurate method of determining the efficiency of the vitamin products. It has been shown by many researchers that considerable variation exists between species and their ability to utilize vitamins.

VITAMIN SUPPLEMENTS

In commercial practice it is vital to ensure the levels of certain vitamins in the diet, and this can only be done by careful calculation at the time the formulations are made.

Though many vitamins will be supplied in adequate amounts by normal feedstuffs, others may be deficient. Unless special provision is made, these will include vitamin A; the following members of the vitamin B complex: choline, niacin, pantothenic acid, riboflavin, vitamin B_{12}; and, occasionally, folacin and vitamins D_3, E and K.

To avoid this danger, it is now standard practice to supplement a food formulation with a specially prepared vitamin concentrate offered by one of the many specialist manufacturers. Thus vitamins are usually included in the form of a pre-mix (or 'spreader' because of the low rate of vitamin inclusion relative to the bulk of the mix). Vitamins A and D_3 are generally supplied together in the proportion of 5 parts of vitamin A to 1 part of vitamin D_3.

These special supplements are not always confined to vitamins. They may be offered in the form of a vitamin/trace mineral mixture, or a protein/vitamin/trace mineral mixture.

The vitamin content of these supplements is expressed in two ways: the requirements for the fat-soluble vitamins are stated in terms of international units per kilo or per pound of mixed food, whereas those for the water-soluble vitamins are usually expressed as mg per kg or per pound of food.

From what has already been said it will be clear that supplements must be stored carefully. Ideally they should be included in a mixture made immediately prior to use. Unfortunately this is not always practicable so that if a mix is made and then, for some reason, a delay occurs before its use, the added vitamins will lose some of their potency. Sometimes an attempt has been made to

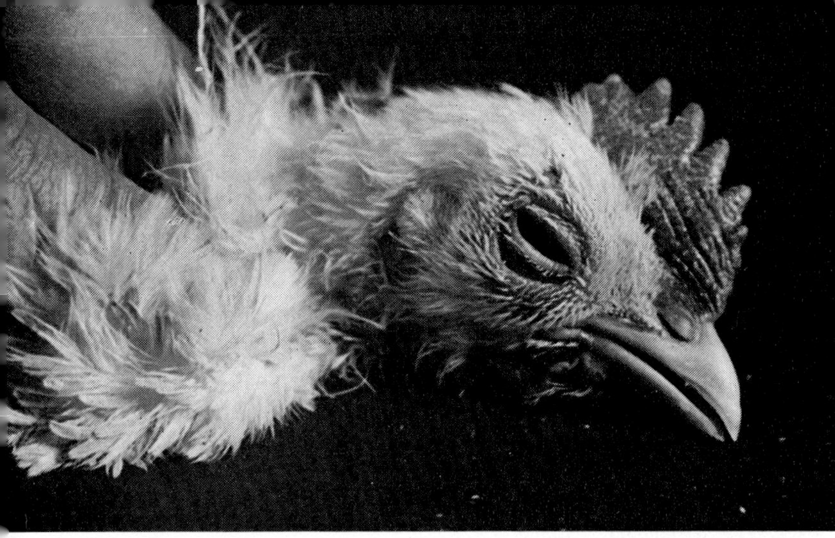

1. Prolonged vitamin A deficiency may result in the ultimate destruction of the eye. *Photo:* Roche

2. Vitamin B1 deficiency may result in the neck becoming twisted due to paralysis of the muscles. *Photo:* Dr C. A. Elvehjem, University of Wisconsin

3. Poor feathering and growth may result from niacin deficiency as in the case of the chick on the right.
Photo: Dr C. A. Elvehjem, University of Wisconsin

4. Emaciation, with the formation of crusty scales at the corners of the mouth and the margins of the eyelids, may be a symptom of pantothenic acid deficiency. *Photo*: Dr Paul Hilbrich, Hanover

5. One of the symptoms of riboflavin deficiency is a tendency for the birds to sit on their hocks with their toes twisted inwards.
Photo: C. R. Grau, University of California

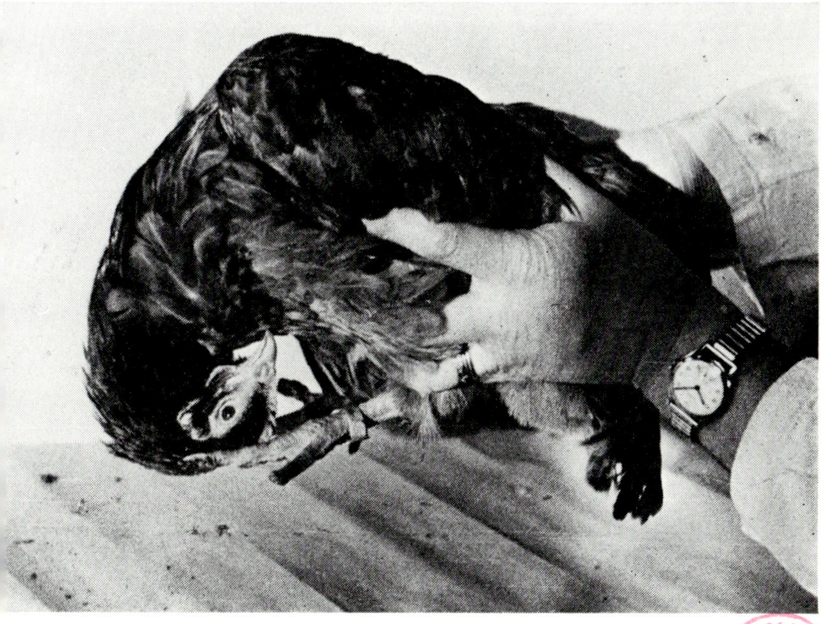

6 The first symptom of vitamin E deficiency is generally a nervous disturbance. Muscle movements are erratic and there may be a backward movement of the head.
Photo: H. Kivimoe, Statens Husognsforok, Uppsala

7. Butyl rubber may be used to construct prefabricated silos for grain storage. In the illustration each silo is designed for 1,000 tonnes of cereals, giving a total of 7,000 tonnes of storage at Banbury, England. The silos are designed for long-term, pest-free, hermetically sealed storage. *Photo:* Butyl Products Ltd

8. One tonne of dry free-flowing material being transported in a mini BULK bag suspended from the tines of a forklift. *Photo:* Pertwee Landforce

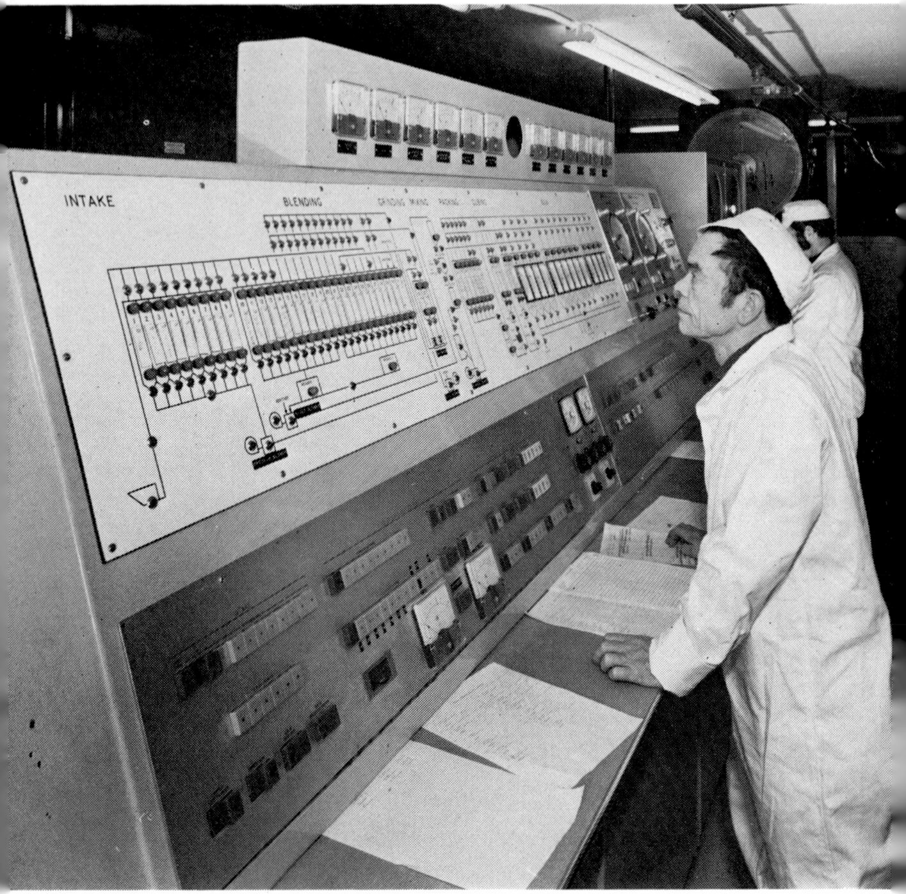

9. The centralized electronically operated control panel for all processes in an animal feed mill. *Photo:* Christy

11. Broiler house prepared ready to receive day old chicks, showing the automatic feeding and water supply system. They are suspended from the ceiling to allow adjustments as the birds grow. The pan feeders are filled by means of a flexible auger and runs of up to 165m may be installed. *Photo:* E. & B. Equipment Ltd

10. Bulk supply bins, each holding up to 11 tonnes of broiler crumbs, arranged to supply food automatically to the feeding system within the poultry house. *Photo:* E. & B. Equipment Ltd

12. A close-up view of a pan feeder showing the electro-controlled wire designed to prevent perching. The pans may be removed for use with chicks from a day old; after a few days the pans are refitted and the system is then fully automatic for poultry of all ages.
Photo: E. & B. Equipment Ltd

13. A circular water trough in which the water is automatically controlled. Troughs of this type are available and may be used by poultry of all species from a day old and onwards. *Photo:* E. & B. Equipment Ltd

guard against this loss during an anticipated delay by increasing the vitamin inclusion by a factor of two or three. If the delay does not then occur it is possible for a ration to cause troubles due to an excess of vitamin intake. This is perhaps most likely in the case of vitamin D_3.

Importance and sources of individual vitamins

FAT-SOLUBLE VITAMINS

Vitamin A

Vitamin A is essential for building up resistance to disease and for the healthy maintenance of the delicate epithelial membranes lining the skin and mucosa of the body. A disease known as nutritional roup may occur in chickens and turkeys when the diet lacks sufficient vitamin A.

Fowls, like other animals, are able to store vitamin A in their liver, and this may be used as a reserve when the diet is deficient. However, chicks have very few reserves on which to call except their unabsorbed yolk. Provided the diet of the breeding hen is adequate, the yolk of the young chicks will also have an ample supply. If the breeder's diet contains insufficient vitamin A, the chicks hatched from the vitamin A-deficient eggs do not get a good start in life (see Plate 1). However, the chick is not completely dependent upon the yolk for it can make use of vitamin A in the diet as soon as it starts to eat food (see Plate 1).

As a fat-soluble vitamin, vitamin A is to be found in reasonable quantities in whole unskimmed milk and in the form of fish liver oils. Unfortunately, the rate of loss by oxidation from the latter products once they are mixed into the food is considerable. On those occasions, therefore, where fish liver oils are used, the oil should be mixed into the diet as near the time of using the food as is practicable. If a starter mash has to be kept for some time, it should be so stored that there is the least possible exposure to the air. If the mixtures have to be stored for three to four weeks, it is advisable to add more fish liver oil. These problems do not arise if a stabilized form of vitamin A is used.

Vitamin A may also be added to the diet in the form of a pro-vitamin. Fresh green foods, like grass, kale, cabbage or lucerne, are a satisfactory source of pro-vitamin A as well as many other nutrients.

When green crops are dried, their vitamin A value may decrease because of oxidation and may vary considerably in the final product. The colour of the product can be used to some extent as a guide to the quality, for samples with a good green colour are frequently also high in pro-vitamin A. Normally, however, a proper analysis is essential. Samples of dried grass and lucerne meal which show a low vitamin A value, may also have a poor content of other nutrients.

The most satisfactory means of supplying vitamin A to poultry is in the form of a dry stabilized product. Together with vitamin D_3, it may be added to the diet in the form of micro-crystalline beadlets made of gelatin. This stabilization of the vitamins helps to guard against the loss of strength. Under normal conditions of storage, potency can be retained for at least six months. Another advantage of supplying vitamin A in this way is the fact that the beadlets are dry, extremely free flowing and, therefore, easy to handle.

Vitamin D_3

Vitamin D is required for bone formation, egg production, reproduction and for the prevention of rickets. A store of the vitamin is passed to the progeny from the breeding hen via the yolk. It is vital, therefore, to ensure that the chicks have the best possible start in life by providing adequate amounts in the diet of the breeding stock.

The importance of the 'carry-over' of vitamin D_3 from the breeding hen to the progeny has been well illustrated by experimental work. For the first two weeks of the chick's life the level of vitamin D_3 in the dam's diet appears to have a major influence on bone formation. After two weeks the vitamin D_3 in the chick's diet has a greater influence. The influence of the vitamin D_3 content of the yolk is felt for the first four weeks of life.

The first symptom of vitamin D_3 deficiency in laying birds is the appearance of thin-shelled eggs, followed by reduced egg production. A very close relationship exists between vitamin D_3 and the absorption of calcium and phosphorus in the diet, the vitamin affecting the rate of utilization of these minerals (see Chapter Five).

Fowls have the ability to build up their own individual requirements for vitamin D when they are exposed to natural sunlight. The sun irradiates sterols lying in or under the skin and causes the formation of vitamin D_3. In areas of low sunshine birds kept extensively, like those kept intensively, will require vitamin D supplements in the diet.

Fish liver oil is another source of vitamin D_3, but the practical problems of adding the oil to food mixtures with the risk of the oil becoming rancid, mitigate against its general use. As previously noted the value of vitamin D_3 is gradually lost once it is mixed in the food. To overcome this problem an extra 50 per cent of the vitamin is usually included in the mixture as a margin of safety. The most convenient means of supplying vitamin D_3, therefore, is by means of a dry vitamin D_3 supplement (see pages 64–65).

Intensively kept birds require a vitamin D supplement in their diet at all times, and it is also a wise precaution to include vitamin D in the diet of extensively kept breeding stock when they are producing hatching eggs.

WATER-SOLUBLE VITAMINS: THE VITAMIN B COMPLEX

The development of intensive methods of production, followed by the demand for carefully balanced diets, has emphasized the importance of the so-called vitamin B complex in poultry nutrition. The constituent vitamins are not related in any chemical sense, but most are active as constituents of enzyme systems and, therefore, have similar functions.

Probably the first dietary deficiency condition in chickens to become known was that of polyneuritis in 1901. By 1926, it had been established that two factors were associated with the prevention of this disease and, by 1927, they were given the names of vitamin B and vitamin G. Later vitamin B became known as B_1 and vitamin G was termed B_2. Progress was rapidly made and other water-soluble vitamins were discovered and named numerically to avoid confusion until suitable individual names could be found. Thus, the B complex was born, which today includes eleven main vitamins.

The numerical classification does lead to some confusion and now that the names are available, it is common practice to refer to the water-soluble vitamins by their appellations rather than by their numbers.

The vitamin B complex is made up of the following:

- Aneurine, thiamin or B_1
- Biotin
- Choline
- Folacin or folic acid
- Inositol
- Niacin
- Pantothenic acid
- Para-aminobenzoic acid
- Pyridoxine or B_6
- Riboflavin or B_2
- Vitamin B_{12} or cobalamine

Some members of the vitamin B complex are to be found well distributed among common poultry foods. Grass and green food generally contain many members of this group of important vitamins. Some, however, are rarely found in significant quantities amongst naturally occurring foods, and special effort has to be made to ensure that they are present in the diet, particularly in the case of intensively kept poultry.

Aneurine (thiamin)

Aneurine is also called vitamin B_1 or thiamin. Special supplementation will generally not be required except in turkey food mixtures.

Aneurine is comparatively stable, is fairly widely distributed in normal poultry foods, and a deficiency is unlikely under normal conditions. Green food is a reasonably good supplier of aneurine, but the richest source normally available to the poultry keeper is unextracted brewer's yeast.

Aneurine is essential for normal growth and digestion, blood formation, and the building up of resistence to disease. It plays a specific part in the metabolism of carbohydrates. It is also required for the normal functioning of the nervous system. A dietary deficiency may precipatate an outbreak of the type of paralysis known as polyneuritis. There may be some small danger of thiamin if the drug Amprolium is being used as a coccidiostat in the diet, for this acts by interfering with the metabolic function of thiamin on the coccidia (see Plate 2).

Biotin

Biotin is present in most food ingredients, but the richest natural source is dried brewer's yeast. Rations for broilers and turkeys may require supplementary biotin.

Biotin is essential for normal growth and reproduction. It is also essential for the prevention of a form of dermatitis, and may play a part in the prevention of some forms of perosis besides being an essential nutrient for hatchability.

A deficiency of biotin has been associated with outbreaks of fatty liver kidney syndrome (FLKS) in commercial broilers and in layers. In those cases where wheat is the main cereal in the broilers' diet, biotin supplementation may be required. The supplementary levels that have been suggested are 0·12 mg per kg of diet for broiler breeders and 0·8 mg per kg of diet for standard broilers

Choline

Choline is the essential nutrient that is particularly necessary for the prevention of perosis. It is reasonably stable, both in its naturally occurring form and as choline chloride, which is used for supplementation. It may, however, undergo oxidation to betaine for metabolic transfer of methyl molecules, especially if there is a marginal supply of methionine. Although, therefore, choline is fairly widely distributed among normal poultry foods, such as milk products, fish meal, liver meal and soya-bean meal, it is generally advisable to use supplementary choline in the form of choline chloride.

There would appear to be a carry-over of choline from the breeding hen to the young chicks via the egg and, for this reason, some breeders include a choline supplement in the breeder's diet, particularly where the breeding stock are kept under completely intensive conditions. The young chick cannot synthesize choline for itself and does not acquire this ability until it is no longer dependent on its yolk sac. Experimental work has shown that some of the functions of choline may be performed by the amino acids methionine and betaine. Neither of these nutrients, however, appears to completely replace the need for choline in the diet, e.g. betaine cannot replace choline in preventing perosis. Betaine is widely distributed in normal poultry food ingredients. Choline may play a role in the prevention of a type of gizzard erosion encountered in young growing turkeys.

Folacin (*folic acid*)

A deficiency of folacin may lead to anaemia, poor growth and perhaps to paralytic symptoms. Folacin also plays a part in feathering, and a deficiency can lead to feather de-pigmentation in birds with coloured feathers. The presence of adequate quantities of folacin in the diet has also been associated with good hatchability, and with the prevention of certain forms of perosis. Lack of folic acid can impair the production of antibodies to certain diseases.

Folacin is well distributed among normal poultry foods. It is found in good-quality dried and fresh green foods, in milk products, dried yeast and liver meal. Supplementary folacin will normally not be required except for turkey breeding stocks.

The importance of the carry-over of folacin from the breeding stock via the egg to the young bird is emphasized, since young poults cannot consume sufficient folic acid, even from well-fortified rations, to make up for any deficiency in the diet fed to their dams.

Inositol

The existence of a factor called inositol has been known since as long ago as 1928. However, there is little evidence regarding its nutritional role, except that it is apparently necessary for normal growth and is found in the red corpuscles of the blood. Ample quantities are apparently present in normal poultry diets, and special consideration is unnecessary.

Niacin (*nicotinic acid*)

Niacin is found in both fresh and dried good-quality green foods, in milk products, liver meal, and also in dried brewer's yeast. The nicotinic acid in cereals is bound and is, probably, not available to chicks and ducklings though again it can be synthesized from the amino acid tryptophan. This means of supply will tend to be blocked off if, under conditions of high production, the supply of tryptophan becomes limited. It is normal, therefore, to supplement poultry diets with this vitamin.

A deficiency of niacin may result in inflammation of the tongue and mouth cavity, a condition known as 'black tongue'. Niacin is also required for good feathering, proper growth, and for the prevention of a type of dermatitis of the feet and skin. A deficiency of niacin has also been associated with a form of leg weakness in turkeys (see Plate 3).

Pantothenic acid

All specialized poultry food mixtures require supplementary pantothenic acid and calcium pantothenate may be used.

Pantothenic acid is found in a wide range of normal poultry foods, but often in bound form and it is present only in very small amounts in cereals. Milk and its by-products, dried yeast, liver meal, and fresh and high-quality dried grass are also good sources.

Pantothenic acid is necessary for the proper maintenance of health, for good growth, and for the prevention of some forms of dermatitis. A deficiency in rations for breeding stock leads to poor hatchability of their eggs. It also depresses antibody production (see Plate 4).

Para-aminobenzoic acid

Para-aminobenzoic acid, which can be isolated from yeast, is apparently present in adequate amounts in normal poultry foods, and special consideration when formulating food mixtures is unnecessary.

Pyridoxine (vitamin B_6)

It has long been considered that pyridoxine deficiency is unlikely to occur under practical poultry farming conditions, for it is widely distributed in the normal food ingredients. Dried brewer's yeast is an especially rich source. However, research now indicates that supplementary pyridoxine may be required in rations for all high-producing stock. The main function of pyridoxine is in regulating the metabolism of amino acids and proteins. With today's use of amino-acid analogues, particularly synthetic methionine, the need for pyridoxine supplementation becomes more evident. The vitamin is also concerned with the function of the adrenal gland which is active in monitoring the 'stress' conditions to which high-producing chickens are frequently exposed.

Pyridoxine, also called vitamin B_6, is essential for the maintenance of good health, growth and production. The vitamin has a sparing action on the amino-acid methionine, and is slightly water soluble.

A deficiency of pyridoxine may result in abnormal excitability, slow growth, reduced egg production and hatchability, and in perosis. In ducks, a deficiency may result in poor feathering.

Riboflavin (vitamin B_2)

Milk and its by-products, liver meal, dried yeast, good-quality condensed fish solubles, and good-quality fresh and dried green food are good sources of riboflavin. However, it is standard practice to include supplementary riboflavin in poultry rations, since relatively high rates of inclusion are regarded as necessary for optimum performance, e.g. 2 mg/kg for layers and 3 mg/kg for breeders.

Riboflavin, also known as vitamin B_2, is one of the vitamins necessary for hatchability and also for growth. A deficiency of riboflavin in the diet of breeding birds will result in reduced hatchability when many embryos may die during the first week of incubation and show maximum mortality at about the eleventh day. A second peak of mortality occurs just prior to hatching, and this results in an abnormal number of 'dead in shell'. When these are examined, a characteristic clubbing of the down may be seen.

The main symptom of deficiency in young fowls is a form of leg weakness known as 'curled toe paralysis'. The birds tend to walk on their hocks, and their toes are seen to be in-turning. An acute dermatitis may also be a sign that chicks are being fed a riboflavin-deficient diet. Riboflavin may play a part in the prevention of perosis, and is also necessary for good feathering (see Plate 5).

Unlike vitamins A and D, riboflavin is very stable, and its value is not easily lost when mixed with other foods. It is water soluble and relatively heat resistant.

Vitamin B_{12}

Good-quality white fish meal, fish solubles and other feedstuffs which are rich in animal protein are usually good sources of vitamin B_{12}, and are the standard method of including this vitamin in the ration. If these products do not constitute 5 per cent of the ration, it is advisable to include supplementary vitamin B_{12} in the mixture. In all cases, turkey foods should generally include a vitamin B_{12} supplement.

The vitamin is essential for good growth and for high hatchability. It is always more necessary in diets which are exceptionally rich in protein or fat.

OTHER VITAMINS

Vitamin C (ascorbic acid)

Chickens and turkeys need vitamin C for their metabolism, but it does not require to be added to the diet. Under most conditions birds of all ages have the ability to synthesize sufficient vitamin C in their bodies to meet their requirements.

There have been suggestions that the supplementation of layers' rations with vitamin C during periods of very hot weather may be an advantage, and further research work is indicated.

Vitamin E

Vitamin E is found in cereal grains, wheat by-products, and in good-quality fresh and dried green foods. Wheat-germ oil is the richest natural source. However, it is advisable to supplement all specialized poultry foods with vitamin E.

Vitamin E is fat soluble and relatively stable, but it may be unstable in the presence of rancid fish oils. It is in fact an antioxidant and is broken down in preventing the oxidation the fat would bring about. It is spared if the ration includes an alternative anti-oxidant.

Vitamin E is essential for the maintenance of good health and for high hatchability. It appears that vitamin E supplementation is particularly helpful in overcoming the end-of-season drop in fertility and hatchability that so often is a feature of turkey breeding.

Adequate levels of vitamin E play a part in the prevention of some forms of enlarged hocks and in the prevention of 'crazy chick disease', which is characterized by a loss of muscular control (see Plate 6).

There is some evidence that vitamin E in vegetable tissue is not so readily available for poultry compared with vitamin E found in oil concentrates. Evidence is further available that some of the effects of vitamin E deficiency can be overcome by selenium, but this is not yet fully understood. The vitamin also shows interrelated effects with the sulphur-bearing amino acids.

Vitamin K

All specialized poultry food mixtures should be supplemented with vitamin K. The inclusion of 2·5 per cent high-quality dried grass or lucerne (alfalfa) will normally supply all the vitamin K required.

The vitamin is passed on to the young chick via the egg and its function is associated with the proper clotting of the blood. A deficiency of vitamin K in the diet may result in an outbreak of haemorrhagic disease amongst broilers and young chicks.

Vitamin K is fat soluble, heat resistant and very stable. Synthetic, water-soluble forms are available and the product most widely used is menadione.

Symptoms of vitamin K deficiency appear to be confined entirely to young growing birds and are not met with in adults. This may perhaps be due to the fact that older birds are able to synthesize sufficient vitamin K in their intestines.

It has been suggested that the inclusion in the feed of some drugs, e.g. sulphonamides, may increase the need for vitamin K. This may be because of the drugs themselves or because of the coccidial infection which they have been introduced to control.

CHAPTER FIVE

Production Regulators: Group Two – minerals

The importance of minerals in poultry nutrition

It is not possible to generalize about the role of mineral ions in poultry nutrition. Some manifestations of the importance of individual minerals have become so well known that they are taken almost for granted. A deficiency of manganese, for example, may be responsible for that bowing of the legs known as 'slipped tendon' or perosis. Again, a deficiency of this same mineral in the diet of the breeding hen may lead to reduced hatchability. More obvious still is the fact that approximately 9 per cent of the egg is made of mineral ions, mainly calcium, while minerals represent over 40 per cent of bones, where they are present mainly as calcium and phosphorous. It is less well known that minerals comprise about 1 per cent of the edible portion of the carcase. In fact, mineral ions, in compound form, are to be found in all body tissues and fluids. Calcium is essential for blood clotting, chlorine is an important constituent of the secretion of the proventriculus; iodine is an essential constituent of thyroxine produced by the thyroid.

The presence or absence of mineral ions – sometimes in small quantities, sometimes in large – is seen, therefore, to be a matter of considerable importance. It is less easy, however, to say with complete accuracy what the requirement of the bird for each mineral may be, for the reasons discussed in this chapter.

CHEMICAL DETERMINATION OF MINERAL CONCENTRATION

The analytical chemist calls the mineral content of a food the 'ash'. This is because when the combustible portion of the feedstuff has been removed, it is the 'ash' or the non-combustible fraction which remains. The 'ash' will generally contain a large number of different minerals.

Table 7 The mineral analysis of a chicken's egg minus the shell (*Poultry and Egg National Board, 1964, USA*)

Nutrients per 100 g	Minerals (ash)	1·0 g
	Calcium	54·0 mg
	Chlorine	100·0 mg
	Copper	0·17 mg
	Fluorine	0·06 mg
	Iodine	12·0 mg
	Iron	2·1 mg
	Magnesium	9·0 mg
	Manganese	0·04 mg
	Phosphorus	210·0 mg
	Potassium	149·0 mg
	Sodium	110·0 mg
	Sulphur	233·0 mg
	Zinc	1·3 mg

The importance of these minerals, whose presence may be determined in the various tissues of the bird's body (as for the egg in Table 7), is somewhat difficult to assess. Sometimes a large amount is needed and is included as a recognizable percentage of the ration; sometimes a minute amount, not more than a 'trace', is essential at an inclusion rate of a few parts per million of the

Table 8 Gross use of mineral ions by avian species

'Macro'-minerals needed on a percentage basis	*Intermediate minerals needed as parts per thousand or per ten thousand*	*'Micro'-minerals needed in parts per million*	*Minerals identified in tissue analysis for which no need is known or satisfactorily established*
Calcium	Iron	Cobalt	Aluminium
Chlorine	Magnesium	Chromium	Arsenic
Phosphorus	Manganese	Copper	Boron
Potassium	Silicon	Fluorine	Bromine
Sodium	Sulphur	Iodine	Cadmium
	Zinc	Molybdenum	Lead
		Selenium	Nickel
		Vanadium	Rubidium
			Strontium
			Titanium

diet. Sometimes minerals are found in the tissues when no dietary requirement for them has yet become apparent. Accordingly the classification shown in Table 8, is made on the basis of the 'gross' use of the mineral by different classes of stock.

Table 9 Classification of minerals in respect of their tolerance by chicken

Mineral	*Requirement ppm*	*Ratio of tolerance levels to requirement*	*Tolerance level ppm*	*Ratio of toxic level to tolerance level*
A. Very low tolerance level				
Cadmium	—	—	0·5	× 10
B. Medium tolerance, relatively high requirement				
Vanadium	3·0	× 5	15·0	× 8
C. Medium tolerance, low requirement				
Arsenic	—		10·0	× 10
Cobalt	0·1	× 60	6·0	× 5
Lead	—		15·0	× 4
Mercury	—		5·0	× 4
Molybdenum	0·24	× 20	5·0	× 20
Selenium	0·1	× 40	4·0	× 2·5
D. High tolerance				
Chromium	0·1	× 1000	100	
Copper	4·0	× 60	250·0	× 2
Fluorine	1·0	× 300	300·0	× 3
Iodine	0·35	× 300	100·0	× 2·5
Iron	80·0	× 20	1600·0	× 1·5
Manganese	55·0	× 10	600·0	× 6
Nickel	0·006	× 10,000	60·0	× 10
Zinc	50·0	× 20	1000·0	× 2

These differences in use are acknowledged in the terminology of the trade. The term 'bulk minerals' is applied by the food compounder to those minerals that are included in the food mixture in relatively large amounts, and include limestone flour, steamed bone flour, di-calcium phosphate, and also salt.

The term 'trace mineral' on the other hand is applied to those minerals that occur in the tissues of animals in very small quantities, and where the amount included or present in the diet is measured, for example, in milligrams per kilo. The term originally arose at the time before accurate methods of calculating the nutrient content of a feedstuff were developed, but when it was important to indicate that a mineral element was required, although at a minute or trace level.

One other point arises in respect of elements found in the tissues in very small quantities. Would larger concentrations be harmful? Do we, in fact, need to know what the tolerance of the fowl may be for some of the minerals quoted, particularly perhaps for arsenic, lead and copper, and is it possible to supply the trace requirement without subjecting the bird to the possibility of poisoning? This proves to be a very important question and it is necessary to know what are the safety margins.

It is obvious that the concentration of any or all of the minerals in the first three categories must be studied with considerable care.

THE MINERAL CONTENT OF POULTRY FOODS

The mineral content of normal feedstuffs varies considerably and whilst those of animal origin are generally well supplied, those of vegetable origin usually have a low proportion.

Animal products used for poultry feeding, and fish meal in particular, generally contain large quantities of minerals. In those cases where, for example, 10 per cent of fish meal is included in the poultry diet, the full requirements of the bird for a number of minerals may be supplied. At the same time, such a food mixture might well contain excessive levels of some other minerals, e.g. salt.

The mineral content of poultry feedstuffs of vegetable origin reflects the minerals present in the soil on which the crops are grown. In those cases where iodine or manganese is deficient in the soil, then the feedstuffs grown on the soil may well be deficient in these important minerals. On the other hand, the feedstuffs may contain toxic levels of certain minerals due to the presence in the soil of comparatively high levels of a mineral, e.g. grains grown on soil with a high level of selenium may contain toxic levels of this mineral.

The development of new food sources and improvements in the method of manufacture or processing of poultry feedstuffs may, from time to time, give rise to changes in the mineral content of the feedstuff concerned, and, unless the appropriate action is taken, problems associated with the relevant mineral will arise.

MINERAL SUPPLEMENTATION

Although a wide range of minerals is found in the feedstuffs normally used for poultry food mixtures, these may not supply the bird's requirements, particularly under intensive conditions. To overcome the possibility of a deficiency mineral supplements are added to the ration.

Considerable care is necessary in calculating the minerals to be used for a particular food mixture and the constitution of the supplement will depend upon the quantity and the availability of the minerals in the ingredients. An exact estimation of the mineral content of the formula has to be calculated, using information from standard tables showing the averages.

It is important to ensure that the diet of the fowl does not contain excessive levels of minerals which could be harmful. Samples of inferior quality meat and bone meal or fish meal are often a source of trouble. Whilst birds try to excrete minerals that are surplus to their requirements, toxic or semi-toxic accumulations may occur leading to serious economic loss in performance.

The presence of certain minerals may interfere with the absorption by the bird of other nutrients. Thus an excessive level of iron, or of elemental sulphur, may interfere with the absorption of vitamin D (unless this has been supplied in protected form), and result in the appearance of a form of leg weakness called rickets. A very high level of sulphur may also hinder the deposition (formation) of yellow pigment in the shanks and the skin.

Similarly, copper sulphate in the diet has been shown to interfere with the prophylactic effect of arsenicals that have been included in the diet for the prevention of blackhead (histomoniasis) in turkeys.

Some minerals may have the effect of binding other mineral ions to themselves, and thus countering some of the harmful effects which the latter might possess. This is called a chelating effect. There are a number of examples of this interaction, including that of arsenic which may counteract the effects of excessive selenium, of copper which may bind excess molybdenum, and of potassium which may provide some protection from the effects of very high sodium levels.

In practice, however, deciding what mineral additions should be made to the ration is an easier task than it may seem. The specific mineral requirements of poultry are to be found in Tables 16 and 17. The rules for their use are simple.

For every formulation the levels of calcium, phosphorus and salt should be calculated separately and specific inclusion of, say, limestone flour is arranged. By contrast the trace minerals are best supplied as a trace-mineral supplement, thus making it unnecessary to consider each of them separately.

The use of 'macro'-minerals

CALCIUM AND PHOSPHORUS

Calcium and phosphorus are considered together because of their close nutritional relationship. The majority of commercial poultry food mixtures require to be supplemented with calcium and many will also require additional phosphorus.

The daily requirements of the laying and breeding hen for calcium are high and vary with the level of egg production, there being an output of at least 2 g of calcium in each egg produced. Probably two-thirds of this is likely to come from ingested calcium; the remainder from body stores. Average requirements and the levels for inclusion in laying foods are dealt with in Chapter Eight.

The most convenient method of supplying calcium is in the form of limestone flour or granular limestone, mixed in the food, or as limestone grit or oyster shell which is usually made available at choice, or supplied in standard quantities at regular intervals.

One of the most satisfactory means of supplying phosphorus is as di-calcium phosphate and this, at the same time, supplies appreciable quantities of calcium.

Research has shown that particle size of calcium carbonate has a significant effect on calcium retention in the gizzard, where large particles can remain for up to four days, during which the calcium is metered out to the digestive system. It would appear that most of the metering-out takes place during the day and very little by night.

In the case of starter and growers' diets it has, in the past, been customary to supply calcium and phosphorus in the proportion of 2:1, the so-called calcium–phosphorus ratio which is the level at which these minerals are found in the bones. As knowledge regarding the requirements for calcium and phosphorus has improved, it has become possible to be more precise in stipulating the amount of these nutrients to be supplied.

The function of calcium

Calcium and phosphorus are vital constituents of the blood and other body fluids and are required in the blood for proper clotting. Both minerals are also intimately linked with bone formation and with egg production.

Immediately prior to laying, calcium is stock-piled in the bone marrow in sufficient quantities for the first six to thirty eggs. Once egg production commences the bird draws on this reserve in the bones at the same time as it utilizes ingested calcium from its layer ration. Thereafter the reserves are replenished on a day-to-day basis, and it is very important that this cycling of calcium through the bone marrow is established and maintained. If the cycle is not induced the bird may lay soft-shelled eggs even when there appears to be sufficient inclusion of calcium in the ration. Naturally the use at this time of a diet that is low in calcium will soon be followed by the production of thin-shelled eggs.

It is a popular misconception that calcium can be stored by the pullet during the rearing period in sufficient quantities to meet the later needs for long-term egg production.

The association between calcium and caged layer fatigue

Brittle or fragile bones may sometimes occur amongst laying

poultry housed intensively. The problem appears to occur more frequently amongst birds housed in battery cages than with birds kept in other systems of housing, and it has been suggested that the placing of pullets in cages may prevent their skeletal systems from developing adequately compared with pullets housed on litter.

It seems more likely, however, that the failure which occurs is in the correct metabolic cycling of calcium through the bone marrow described above. Low-calcium diets fed just before caging may help the bird's parathyroid gland to initiate this cycle properly.

Bone fragility also appears to develop with age and can become a problem at the time of processing, where, unless they are sufficiently hard, breakage or shattering of the bones may occur. If calcium uptake has proceeded normally this should not occur.

Calcium excess

Whilst the importance of calcium can be gauged from the facts that the shell of the egg is almost entirely composed of calcium carbonate, and that the thickness and the strength of the egg shell is linked with the presence in the diet of adequate calcium, phosphorus and other nutrients, it is also true that the levels of calcium in the diet to give maximum shell thickness may be too high to give the maximum rate of egg production. Calcium levels can be too high and not fully productive.

The requirement for calcium and phosphorus in the diet is influenced by the amount of vitamin D present. It has been shown, for example, that when the diet is low in vitamin D, the requirements for these two important minerals may increase. The presence of vitamin D in the food is necessary to allow the bird to make use of dietary calcium. Thus when a diet is low in vitamin D, the absorption by the bird of calcium may be reduced, thus leaving an excess of calcium in the digestive system which may then combine with phosphorus, and be rendered unavailable to the bird (see page 83).

Calcium may have an antagonistic effect on some antibiotics, e.g. tetracyclines. In some cases, when the tetracyclines are used for veterinary purposes it may be considered necessary to achieve a high level of the antibiotic in the blood. To accomplish this a low-calcium diet may be recommended. Low-calcium diets should only be used for limited periods under veterinary advice.

An excess of calcium or phosphorus may easily occur under practical conditions. The ration may already contain adequate quantities of these minerals, but the balance can be upset by a supplementary mineral mixture, added in the mistaken belief that this may be of value. Another possible cause of an excess of calcium lies in its relative cheapness, for there is a danger that an additional amount may be used in order to lower the cost of the food mixture.

The need for phosphorus

The value of phosphorus in poultry nutrition is dependent upon the form in which it is present in the diet. Phosphorus from cereal sources is in the form of phytates and is relatively unavailable to poultry. For optimum results it appears that at least 0·4 to 0·5 per cent of the total phosphorus in the diet, and probably all of it, should be from non-cereal sources. Inorganic phosphorus is much more readily available and good sources are fish meal, meat and bone meal, steamed bone flour, and anhydrous di-calcium phosphate.

The term 'available phosphorus' is used to refer to the inorganic phosphorus. All the phosphorus from animal sources, and a maximum of 40 per cent of that from cereal and vegetable protein sources, may be regarded as being 'available phosphorus' in the formulation of poultry diets.

An excess of phosphorus is often found to be due to the use of a very large quantity of fish meal or meat and bone meal and may lead to the poultry showing symptoms of manganese deficiency.

SALT

Salt is made up of two minerals, sodium and chlorine, both of which are essential for proper metabolism, for growth, digestion, blood formation, and the maintenance of the acid/base balance.

Food ingredients rich in animal protein contain generous quantities of salt, whereas those of vegetable origin contain little. In consequence the inclusion rate for salt in a food mixture must depend upon the salt content of the other ingredients already specified. In some cases little or no additional salt will be required. Generally, however, the inclusion rate will range from 0·25 to 0·3 per cent corresponding to 0·1 to 0·15 per cent sodium. Adult chicken can tolerate much higher inclusions of salt in the diet (by a factor of about 10) but this leads to excessive water consumption, loose droppings and reduced production. Turkeys are even more vulnerable to this so-called salt poisoning.

Good-quality agricultural feeding salt, or iodized salt, will be suitable for use as a supplement.

Too much salt in the drinking water is more damaging than an excess in the ration. One per cent of salt by volume in the water is as harmful as 4 per cent by weight in the ration. 0·5 per cent is suggested as the upper desirable limit for the salinity of drinking water. This may be a point of particular importance in relation to ground water in an area intended for poultry development.

POTASSIUM

Potassium is an essential mineral and is required, together with sodium and chlorine, for the proper maintenance of the acid/base balance.

Adequate amounts of potassium are generally supplied by normal poultry feedstuffs, and special supplementation is usually unnecessary. Alfalfa meal, molasses, and soya-bean meal are good natural sources of potassium.

'Micro' – and trace minerals

COBALT

Cobalt is an essential mineral that is found as a constituent of vitamin B_{12}. There is little chance of a deficiency occurring with normal poultry food ingredients, for cobalt will be supplied by the vitamin B_{12} present naturally in the diet and by the B_{12} supplement that is used.

The value of cobalt appears to be entirely concerned with vitamin B_{12}, which is particularly vital to the circulatory system.

COPPER

Copper is an essential nutrient, but its precise nutritional role in poultry metabolism has not been clearly defined. Most specialist poultry food mixtures should contain a copper supplement, especially if the diet has a high content of zinc.

Copper is apparently essential for the highest levels of hatchability, for helping to build up resistance to disease, and for the development of blood pigments. Copper is also needed for the proper utilization of iron in the development of haemoglobin.

Copper oxide, sulphate or hydroxide may be used for supplementing the food mixture. Copper compounds may be contaminated with arsenic which should not be allowed to exceed 40 ppm of the final mixture.

FLUORINE

Fluorine is a mineral that may be essential for poultry nutrition, possibly for bone formation, but a firm basis for its use has not yet been established. Fluorine is widely distributed in normal poultry foods and a deficiency is unlikely to occur.

On the other hand, some samples of phosphorous products may contain levels of fluorine that could be toxic if used in poultry diets. This is one of the reasons why unusual sources of phosphorus should not be used until after careful testing, including feeding tests with chicken. The maximum levels of fluorine in di-calcium phosphate should be 0·2 per cent.

IODINE

Iodine is an essential constituent of thyroxine, the hormone secreted by the thyroid. Eggs produced by birds fed an iodine-deficient diet will give poor hatchability.

Iodine is normally supplied as part of the trace mineral mixture, included in most poultry rations at the rate of 0·5 ppm.

Fish meal is usually considered as a good source of iodine, and where this forms a relatively large proportion of the diet (e.g. 5 to

10 per cent) the addition of an iodine supplement will not be necessary.

Food ingredients of vegetable origin are generally very low in iodine, and where these are used to provide the energy and the protein of the diet, an iodine supplement should be added to the ration. Also, if rapeseed meal is used as an important source of protein, iodine will be needed to counteract the goitrogens contained in this meal.

The need for iodine may be increased when the diet contains a high percentage of calcium and phosphorus. It has been shown that a high level of iodine may be detrimental to egg production, and may affect semen production and fertility.

The following products may, after stabilization, be used for providing supplementary iodine. All are white crystalline powders.

	Level of iodine per cent
Calcium iodate	65
Potassium iodate	59
Potassium iodide	76·5

IRON

Iron is an essential constituent of the haemoglobin of the blood, but requires the presence of copper for its proper utilization.

It is the normal practice to supplement specialist poultry foods with iron although the precise basis for so doing is not clearly defined. The normal sources include iron carbonates, oxides and sulphates; all are very insoluble and absorption rates in adult birds are as low as 5·5 to 10 per cent. Iron does not seem to be a cause of toxicity.

MAGNESIUM

Magnesium is an essential nutrient but no supplementary magnesium is usually necessary in specialist poultry food mixtures, because conventional foods contain adequate quantities of the mineral.

High levels of magnesium in the diet of chickens have been shown to act as a laxative that reduces the time required for the passage of food through the digestive system. It is for this reason that the use of dolomitic limestone, which contains magnesium, should be avoided. Magnesium is otherwise non-toxic; thus the 'toxic' level of magnesium for young chickens would appear to be over 6,000 mg per kg of diet. Turkey poults have a relatively high requirement of 450–600 ppm of magnesium to total ration. This figure is comparable to those for egg-laying hens in full production.

MANGANESE

Manganese is an essential nutrient that is required for bone formation, the prevention of some forms of leg weakness (perosis), for optimal growth, for egg production and for high hatchability. A free-flowing manganese supplement should normally be included in all poultry foods.

Commercial powdered manganese carbonate, oxide or sulphate may be used (all are relatively stable) and have the following average manganese content:

	Manganese per cent	*Colour*
Oxide	60–62	Greenish-black or brown
Carbonate	40–50	Brown to black
Sulphate	31–32	White to pale pink

The actual manganese content of the product used for supplementing the food mixture should be known before it is used, for a wide range in potency occurs in the different forms of manganese compounds. In the presence of high levels of phosphorus additional manganese is required.

Manganese is found in a large number of normal feedstuffs, but frequently this source is insufficient for satisfactory results. Good-quality millers' offals are a particularly rich source of manganese which is contained in the outer covering of the wheat germ.

MOLYBDENUM

Molybdenum is a mineral which perhaps may be essential for correct metabolism, but a proper basis for its need has not yet been established. Special consideration need not usually be given to molybdenum when formulating poultry foods since, if it is required, it is in very minute amounts. It has been shown to be a component of some enzymes, and has recently been used in breeder diets to overcome reduced hatchability and a high level of cull chicks.

Copper has been reported as being antagonistic to molybdenum and may offset potential excess of the latter. Thus aortic rupture in birds fed on a diet that is low in copper, has been ascribed to a toxic level of molybdenum.

Two products (both of which are white crystalline powders) used to supply molybdenum, are:

	Molybdenum per cent
Molybdic acid	68
Sodium molybdate	40

SELENIUM

Selenium is an essential trace mineral and is readily absorbed by the chick; however its precise nutritional function has not yet been fully worked out. There is a close relationship between selenium and vitamin E in the prevention of exudative diathesis in chicks and muscular dystrophy in poultry and chicks. Many of the functions of selenium appear to be performed equally well by vitamin E except that atrophy of the pancreas may occur in the absence of selenium and this cannot be prevented by vitamin E.

Supplementary selenium will rarely be required in rations made from normal poultry food ingredients.

The selenium content of feedstuffs varies considerably and is dependent upon the selenium content of the soil on which the food is grown. One result of this is that grains grown on seleniferous soils may contain toxic levels of this mineral. Grains containing levels of selenium of up to 14 ppm have been reported, whereas it is considered that the minimum toxic level may be as low as 5 to 8 ppm.

Fish meals and fish solubles are probably the richest natural source of selenium. Sodium selenite, containing 45·6 per cent selenium, may be used as a supplement and is a pale brown crystalline powder.

VANADIUM

Vanadium is not generally regarded as an essential mineral for poultry, and it has been shown to reduce significantly the feed consumption and weight gain of growing chickens when it occurs in the diet at a level of 10 ppm. For laying birds, there is some evidence that vanadium has an adverse influence on shell quality, and brittle eggs have been reported when layers were fed diets containing 8 ppm.

Vanadium can be present in toxic amounts, especially in some phosphorus compounds, and such compounds should not be used for poultry feeding.

ZINC

Zinc is an essential trace mineral and is required for good hatchability, growth, bone formation and feathering. Specialized poultry rations should normally contain a zinc supplement.

There appears to be a relationship between the requirements for zinc and the calcium and phosphorus levels in the diet, and high levels of these minerals appear to reduce the availability of zinc, thus increasing the dietary levels required.

Zinc is widely distributed amongst normal poultry feedstuffs, and at one time it was thought that the naturally occurring zinc, plus that obtained from contamination from zinc food and water receptacles, were sufficient to meet each bird's requirements. However, it is now clear that all the zinc from normal feedstuffs may not be available to the bird, and therefore the zinc content of these foods should be ignored when formulating poultry diets. Most poultry food mixtures contain some zinc, and the addition of 50 ppm of a suitable supplement is all that is necessary.

A number of zinc-containing compounds are available including:

	Zinc per cent
Zinc oxide	80
Zinc carbonate	55–56
Zinc sulphate	36

Zinc compounds may be contaminated with lead and the level of zinc itself should not exceed 1,000 ppm.

Minerals with high potential toxicity

ARSENIC

Some organic arsenic compounds, commonly called arsenicals, are sometimes included in the diet of broiler chickens because of their growth-promoting properties, and also to improve pigmentation (see Chapter Six, pages 97–98). Arsenical compounds are also used very successfully, as an aid to good management, for the prevention of some poultry diseases (such as blackhead in turkeys), although this function may be impaired when copper sulphate has also been included in the diet.

Table 10 Arsenicals – levels of use. (Based on information from the *Feed Additive Compendium*, 1975)

Product	*Stock*	*Level of use per cent*	*Application*
Arsanilic acid and Sodium arsanilate	Chickens Layers Turkeys	0·005–0·1	Growth promotion Improved food efficiency Pigmentation Improved egg production
Carbarsone	Turkeys	0·025–0·0375	Aid in prevention of blackhead
Nitarsone	Chickens Turkeys	0·01875	Aid in prevention of blackhead Dangerous if used by ducks and geese
Roxarsone	Chickens Layers Turkeys	0·0025–0·005	Growth promotion Improved food efficiency Pigmentation Improved egg production

The level normally used for growth-stimulating purposes for chickens and turkeys varies according to the compound to be employed, and it is wise to use one where the toxic level is many times that of the recommended inclusion rate. The toxic level of available compounds may range from only twice to nine or ten times that of the inclusion level. If arsenicals are already in use as growth stimulants, their use for the treatment or control of disease should be carried out only under veterinary advice.

Serious losses have occurred following the inclusion of excessive levels of arsenicals in formulation. It is essential, therefore, to calculate the quantities to be used with very great care.

Table 10 shows the arsenicals that have been used in poultry nutrition together with the level of inclusion approved in the USA.

One of the problems concerning the use of arsenical compounds in poultry foods is that traces of arsenic can be deposited in the edible tissue of the chicken or turkey. To overcome this problem the use of foods containing arsenic compounds should be discontinued five days before the birds are slaughtered for human consumption. For this reason the use of arsenicals in poultry foods is restricted by official regulations in some countries.

CHAPTER SIX

'Upper-cylinder lubrication' – additives to the ration

Introduction

The term 'additive' is applied by the food compounder, in a broad sense, to all products other than those commonly called feedstuffs, that may be added to the ration with the object of obtaining some special effect.

Additives are usually included in the food mixture in very small quantities and require very careful weighing, handling and mixing. Those of special interest in the formulation of poultry foods include amino acids, antioxidants, antibiotics, drugs to prevent or control disease, pigments, trace minerals and vitamins.

Amino acids, trace minerals and vitamins have already been considered. The remainder may, for convenience, be considered as falling into two groups. The first group comprises those additives that have a specific nutritional role, and includes fifteen or more growth-promoting substances alone. The second group covers those compounds concerned with the prevention and control of disease and, here, the number used has so far topped sixty. Antibiotics may be included in both groups. The nutritional additives only will be considered in detail and these are shown in Table 11.

The journal *Feedstuffs* publishes regular lists of the different

Table 11

Antibiotics	Hormone-like products
Antioxidants	Nitrovin
Amino acids	Pigments
Arsenicals	Trace minerals
Enzymes	Tranquillizers
Flavouring agents and spices	Vitamins
Hardening agents	Propionic acid

additives available for inclusion in animal foods together with details of the pre-slaughter withdrawal times (i.e. the time before the slaughter of the animal when the ingredient should be withdrawn from its ration to avoid the possible danger that traces of the ingredient might appear in edible tissues and hence in food offered for human consumption).

These 'nutritional' additives do not have a part to play in the metabolic processes outlined in Chapter One. What they do is to make the processes more effective in some way, either by counteracting some growth depressant in the system, by modifying hormonal balances or by improving the appeal or quality of the food mixture on offer and thereby the birds' consumption of it.

Obviously the use of additives should only be considered when there is evidence that their inclusion in the mixture will lead to some such advantage and that the advantage will be economical. Their use should be very carefully controlled at all times in case there should be effects which will be counter-productive.

In the case of broilers where high liveweight gain is at a premium a suitable growth promotant is now included in the diet as a matter of routine.

METHOD OF USE

Additives are included in a food mixture in very small quantities and, in view of this, special consideration must be given to the manner in which they are mixed, in order to ensure that the additive is evenly distributed throughout the mixture.

The most satisfactory method of mixing is to incorporate the additives in a 'pre-mix' or 'carrier' before they are added to the main bulk of the ingredients in the mixer. The pre-mix might 'bulk up' the additive in this way so that it may be added at the rate of, say, 3 kg per tonne of ration.

Considerable care is necessary when calculating the actual quantity of the additive that has to be included in the ration. The inclusion of incorrect amounts may lead to unexpected, perhaps, unfortunate results.

In some cases there can be a danger to public health. There may, with some additives, be a build-up of a specific mineral or compound, in the edible tissues of the bird, or in the egg. To guard against this possibility, a number of countries have regulations limiting or prohibiting the inclusion of additives in animal feedstuffs. Inclusions must be detailed on a label supplied to the customer along with the food container. In addition, the manufacturer should give details of any pre-slaughter time during which the drug must be withdrawn from the diet. Where this is necessary the time is usually a period of three to five days.

The poultry farmer who buys food in compound form requires to know the types and the inclusion rates of any additives in the ration that he has purchased. It is also necessary for him to know whether the use of the product requires any special safety precautions, either to protect the attendants handling the food or the stock which may consume the fortified mixture. If he is mixing a purchased concentrate with home-grown ingredients; he must know if the concentrate will cause contamination of the mixing equipment, which could have unfortunate results on other stock consuming diets subsequently mixed in the equipment.

The veterinarian responsible for the disease-prevention programme of the stock also requires to know the type and levels of additives used.

The most practical method of supplying the information required, so that it is available when it is needed, is to attach a label to every bag or carton of the ration or, in the case of food delivered in bulk, to include the information on the delivery ticket.

The label should include the following information: the short chemical and common name of the additives, the levels of inclusion and any special instructions regarding the manner of their use. The latter may include, if necessary, a warning against the use of the supplemented food for certain classes of stock – for example, breeding stock – or a warning that other additives should not be used at the same time except as recommended by a veterinarian.

The possible effect of one additive upon another must always be considered. In some cases additives may be antagonistic one to another whereas, in others, they may be complementary. The combined value of two additives used at the same time may on occasion be greater than the sum of their individual values, and this phenomenon is known as a synergistic effect.

Additives intended to overcome some limit to growth

ANTIBIOTICS

Antibiotics may be used for improving rate of growth and food efficiency as well as for the prevention and treatment of disease. The so-called nutritional levels of inclusion are 4 g per tonne of the narrow-spectrum antibiotics, and 10 g per tonne of the broad-spectrum types.

High prophylactic levels of inclusion would be at the rate of 50 to 100 g per tonne for seven to ten days, followed by 25 to 30 g per tonne for a further ten days, or until the antibiotic is no longer required. High-level antibiotics should be used only on the advice of the veterinary surgeon but inclusion in the ration is a convenient way of ensuring that all birds in a flock are treated evenly.

An antibiotic has been described as a substance produced by one micro-organism that is antagonistic to the growth of others. This definition may give a clue to the reasons why the use of antibiotics in poultry food mixtures may sometimes lead to better results. Some antibiotics appear to be selective in their action in that they affect specific organisms without apparently influencing others. In such cases, they might destroy harmful pathogenic bacteria without harming useful bacteria involved in the processes of digestion. As a result, a change in the balance of organisms that inhabit the bird's intestines is brought about and often this may help the bird to make better use of its food.

Suitable antibiotics, it may therefore be argued, could neutralize any unsuitable bacteria or other toxic organisms, thus rendering further nutrients available and, at the same time, stimulate some of the useful intestinal micro-organisms thus allowing the greater production of certain nutrients.

There is some evidence to suggest that the effect of antibiotics may be due to the presence in established poultry units of a low level of infection by certain disease micro-organisms which the antibiotics overcome. This same effect is usually conspicuously absent if antibiotic-enriched diets are fed to birds in a new poultry house in which no such low-level infection could have built up.

Antibiotics are usually classified as either narrow spectrum antibiotics, which include procaine benzyl penicillin, and those with a broad spectrum of activity, including the tetracyclines, (aureomycin and terramycin), bacitracin, oleandomycin and virginiamycin.

The names 'broad'- and 'narrow'-spectrum antibiotics distinguish between those that are effective against a wide range of bacteria and those which operate against only a few species.

Both narrow- and broad-spectrum antibiotics appear to be equally satisfactory for young birds, but those of broad spectrum are generally used for adults. Antibiotics may be included in either the food or the drinking water, but their use at the 'nutritional' level is generally confined to inclusion in the food mixture.

Claims made for the use of antibiotics in poultry food mixtures include the maintenance of food intake under conditions of stress which, in turn, helps to reduce the loss from culls. More obviously, growth rate is thought to be improved by 5 to 7 per cent.

The use of antibiotics to overcome sub-clinical disease should only be resorted to when management and all other relevant details have been carefully checked; antibiotics are no cure for bad management.

The use of antibiotics in animal foods is covered by Government regulations in the United Kingdom, as indicated on pages 93–94.

There is some evidence that tetracyclines may be more effective in food mixtures that are low in calcium compared with the standard recommendation. The antibiotic is said to be potentiated by the use of low-calcium diets. It must be stressed that any departure from the recommended requirements should only be made following veterinary advice. It is not desirable to lower the calcium levels, or the levels of any other nutrients, below those recommended for a period of more than one or two days, otherwise less satisfactory results may occur.

One method of using antibiotics which attempts to minimize the problem has been to remove all the food from the birds and then to provide a low-calcium diet, together with the antibiotic, for a period during which the birds can consume up to 20 per cent of their normal intake of food for the day. Once the potentiated food has been consumed then normal feeding is resumed.

ENZYMES (see Chapter One, pages 26–28)

Enzymes are complex organic proteins, secreted in the digestive tract, that play an essential role in the chemical changes involved in metabolism without themselves being used up in the process. Only very small quantities of these catalysts are required for satisfactory metabolism. Enzymes occur naturally in feedstuffs and special supplementation of the food mixture is not usually necessary. However, a wide range of enzymes exists and each one has a particular part to play, so it is not surprising that mixtures of enzymes have been included in poultry diets in an attempt to ensure complete digestion of the ration.

Enzyme supplementation of food mixtures containing barley and other cereals, including oats, rye and wheat (but not maize and cereal by-products like millers' offals), has been shown under some circumstances to be capable of improving their utilization and to give drier droppings from the birds.

The supplements used contain a very large number of enzymes, and the one specifically responsible for the improvement in results has not been identified. Response to enzyme supplementation is by no means uniform and is possibly determined by the climatic conditions at the time of harvest of the cereals used in the ration.

Thus poultry fed on diets containing cereals grown in the western parts of the USA respond to enzyme supplements. The same is true of barley grown in the Lebanon. It is less true for barley grown in North Eastern USA or North West Europe. It has been suggested that harvesting during wet climatic conditions may activate the naturally occurring enzymes in the barley and, therefore, the addition of enzyme supplements becomes unnecessary.

ARSENICALS (see Chapter Five, pages 90–91)

Additives under this heading have been widely used in rations for broiler chickens.

NITROVIN

Nitrovin, which is sold under the trade name of Payzone, is a very popular growth promotant used extensively in broiler production.

The claims made for this compound is that compared with controls that do not receive Nitrovin, 5 per cent heavier broilers will be produced after consuming 3 per cent less food.

Nitrovin is used at the rate of 0·5 kg of supplement per tonne of food.

Additives intended to modify the hormonal balance

HORMONES

Hormones are substances produced by the endocrine or ductless glands of the body. They pass directly into the bloodstream and play a vital part in the well-being of poultry of all species, being concerned as regulators in most body processes. Thus growth, maturity, egg production and broodiness are all 'controlled' by hormonal secretions. The presence of many of these in the bloodstream can now be measured and their effects stated in quantitative terms.

The pituitary or hypophysis, situated in the brain, is one of the most important of these glands and produces several different hormones, some of which act as regulators in the growth and activity of the other endocrine organs like the thyroids, the parathyroids and, especially, the gonads.

The gonads produce the hormones which control the sex characteristics of the bird. In the case of the female these are known as oestrogens, or female sex hormones, whereas the corresponding sex hormones in the male are androgens. Under the influence of androgens cockerels and turkey cocks continue to grow lean and active until they have passed the weight at which it is normally required that a well-finished carcase should be sold. If they are 'de-sexed' they will achieve this carcase 'finish' earlier and more economically in respect of food utilization. How, then, can this be achieved?

A number of synthetic compounds have been prepared which have characteristics very similar to the natural female sex hormones. These include diethyl-stilboestrol, dienoestrol-diacetate, hexoestrol, dianisyl-hexene and thiouracil. All are available in the form of tablets, powders or paste. As a result synthetic hormones are often 'implanted' by injection into the bloodstream of growing male chickens and turkeys for the production of 'capons', with the object of improving the market appearance.

Other synthetic hormones have proved useful in suppressing broodiness.

Implantation is usually done by introducing a tablet into a small incision made in the neck at the side of the bird's head, i.e. a portion of the bird that is not usually eaten. The skin is lifted with one pair of tweezers and the tablet is inserted with another. Alternatively hormone paste may be injected with a 'paste-gun', or heavy syringe, which does not require an incision and is easier to use in field conditions. Implanted in this way the hormone is released slowly and regularly into the bloodstream and regulates growth in the desired way. It is deposited in only neglible amounts in the body tissues themselves.

Fowls and turkeys treated with oestrogens become more docile and may develop a somewhat anaemic appearance. On killing, their final appearance and quality are improved, largely owing to the better distribution of the subcutaneous fat. The results of implantation can generally be seen within about seven days.

The hormone implantation is effective for from three to eight weeks according to the preparation used, and should, therefore, be carried out three to eight weeks prior to the date on which the birds are to be killed. The amount used should be that recommended by the manufacturers, whose instructions should be carefully followed.

The effects of hormone treatment are temporary. Males that have been caponized will return to normal activity after the effects of the hormone treatment have passed. However, it is not advisable to caponize potential breeding stock as possible side effects are not fully understood.

There are certain dangers associated with the use of hormones and therefore careful handling by the poultryman is essential. Some individuals may be allergic to the products and may develop a rash on the face. Hormone-like products should be kept under lock and key on the farm and well away from the reach of children.

As pointed out in Chapter Six, pages 93–94, the use of synthetic hormones for animals is often subject to official regulations. The manufacturer's recommendations concerning pre-slaughter withdrawal should be followed exactly.

TRANQUILLIZERS

Some birds fail to grow normally because they are hyper-active, always moving and reacting sharply to noise and movement in the poultry house. The question then arises: Can this excess activity be modified?

Tranquillizers, as the name implies, are a group of substances that have been used in human medicine for reducing symptoms of stress and anxiety. Experiments are being conducted to determine their possible value in poultry nutrition.

The drug commonly used is known as reserpine and has, under experimental conditions, been used at the rate of 0·5 per cent – 500 mg/kg of diet. It may also have some value in the treatment of aortic rupture. The use of reserpine in the diet of breeding turkeys at the time of egg production has been associated with reduced fertility.

The main claim made for the use of tranquillizers is that the drug enables the bird to make better use of its food, thus leading to an improved growth rate.

There is some evidence to suggest that the use of the tranquillizer methimazole in conjunction with stilboestrol may be of value in the chemical caponization of chickens and turkeys.

Tranquillizers used under experimental conditions with poultry include: benactyzine, carbronal, chloropromazine, hydroxyzine, meprobamate, methimazole, perphenazine, promazine, reserpine and serpasil.

Additives intended to improve the acceptability of the ration

ANTIOXIDANTS

All vitamin/trace mineral mixtures, protein/trace mineral/vitamin mixtures, and fats, should contain an antioxidant, e.g. BHT (see Table 12), normally at a level of 0·02 per cent. The antioxidant should be incorporated in the food ingredients as soon after preparation as possible.

Antioxidants increase the stability and storage life of certain nutrients, following their manufacture or preparation for use, which are prone to loss of value due to oxidation.

Antioxidants are used to give protection to fats and oils which might become rancid in the feed mixture, to fat-soluble vitamins which would break down and lose their potency and to the xanthophylls (used for improving the colour of egg yolks and broiler flesh), which might lose their pigmenting value.

The antioxidants may be added to the food mixture itself, or to the pre-mixes. The final food mixture may, without naming them, contain a number of different antioxidants.

Table 12 The antioxidants commonly used in poultry feeding

Butylated hydroxanisole	(BHA)
Butylated hydroxtoluene	(BHT)
Ethoxyquin	(EQ)

Of those antioxidants commonly used BHT and EQ have both been shown to be effective in protecting fat compounds in food mixtures stored for a period of up to twenty-seven weeks at temperatures ranging from 18° C to 26° C.

EQ at 0·125 per cent of the feed mixture may be used to improve the stability of xanthophylls, and thus improve yolk pigmentation. Experimental work has shown that in food mixtures supplemented with EQ the pigmentation value of the xanthophylls was not lost after storage for three weeks at 47° C.

In the diet of fattening poultry prior to slaughter, antioxidants have been used with the object of improving the stability of the carcase fat and reducing its tendency to go rancid. The use of vitamin E together with BHT and EQ has given encouraging results.

FLAVOURING AGENTS AND SPICES

The addition of flavouring agents or spices to poultry food mixtures has not yet been shown to be necessary. Such products do not contribute to the value of the food in any way.

Chickens possess a sense of taste but this differs from that of man. Efforts to find flavouring agents that will increase food consumption have been made, but so far have proved unsusuccessful. Research on this subject will no doubt continue, since the discovery of a product that would lead to improved food intake could be of considerable value to the poultry industry.

PELLET-HARDENING AGENTS

Pellets used for poultry diets should be firm and free from dust. Soft pellets that crumble easily often clog feeders with the result that the birds may not be able to eat the quantity of food they require for maximum growth or egg production. Poor-quality pellets may also lead to waste of feed. Products used to increase the firmness of pellets are known as hardening agents or binding agents.

Molasses at the rate of 2·5 per cent of the mixture is effective as a binder. Sodium bentonite is a popular hardening agent and is used at the rate of 10 to 20 kg per tonne of the ration. Products are available which combine the binding and sweetening properties of condensed molasses solubles together with a hardening agent.

Sodium bentonite has been found to absorb the drug amprolium to a degree which affects chemical analysis. However, this may not necessarily reduce the biological efficiency of the coccidiostat.

PROPIONIC ACID

Propionic acid or its salts are used to control the infestation by weevils, mites and moulds of food ingredients during storage, particularly grains of high moisture content. The preservative is used at the rate of 250 to 500 g per tonne of grain.

Additives intended to improve the appearance of the final product

PIGMENTS

Pigments may be added to the rations for broiler chickens and to those for laying hens when a deeply pigmented skin in the broiler, or richly coloured egg yolks from the layer, are required. This will be most likely when food mixtures low in maize or green food (grass or lucerne) have to be used.

Colouring in naturally yellow-skinned birds is dependent upon the level of deposit in the fat, skin, shank and beak of pigments known as xanthophylls. The majority of strains of broiler chickens are yellow-skinned. In some areas the demand is for well-pigmented birds whereas in others the preference is for white-skinned birds. In areas where white-skinned broilers are preferred this can be achieved by ensuring that yellow maize and its by-products and fresh or dried green foods are not included in their diet during the last four weeks prior to slaughter.

In a similar manner the degree of pigmentation of the yolks of eggs laid by all species of fowls is dependent upon the level of carotenoid pigments in the diet.

The demand from the shell-egg market in respect of yolk colour varies. In the United Kingdom deeply coloured egg yolks are preferred, whereas in the United States a paler yolk is demanded. However, from the liquid- and dried-egg market the demand is for deeply coloured egg yolks which in turn will impart the right degree of colour to the cakes and other products made by the caterers. Eggs with almost white yolks may be produced by eliminating xanthophylls from the diet.

Fresh and dried green food are the main natural sources of xanthophylls and, in addition, carry quantities of the carotenoid precursors of vitamin A. In view of this it is often considered that dark-yolked eggs are also rich in vitamin A, and pale-yolked eggs are low, but this may not be the case. Vitamin A itself is an almost colourless substance and although dark-coloured yolks may also be rich in vitamin A they are not necessarily so.

Maize gluten is another good source of xanthophylls, but is not generally available in sufficient quantities for wide use.

A rapid oxidation loss of xanthophylls from ingredients takes place, particularly during hot weather, but the presence in the mixture of a suitable antioxidant can be of considerable help in slowing down the rate of loss.

The analysis of food ingredients to determine their total xanthophyll content includes both the pigmenting and the non-pigmenting xanthophylls, and the content of the latter may range from 15 to 25 per cent. Table 13 shows representative values for the commonly used pigmenting food in which the total amount of xanthophylls has been reduced by 20 per cent.

Table 13 Content of active pigmenting xanthophylls (i.e. pigmenting for egg yolks) (N. Comben, 1970)

	mg/kg
Grass meals	750
Maize-yellow (US)	12–15
Maize-plate (Argentine)	25–28
Maize-gluten meal, 60 per cent	150

Specially prepared sources of pigments are available for inclusion in the diet. When these are included to supplement the naturally occurring xanthophylls, to provide a level of 10 to 15 mg/kg of pigmenting xanthophylls in the total diet of layers, the yolk colour will normally be acceptable. In the case of broilers the total level of xanthophylls required may reach 22 mg/kg of ration, but the supplementary xanthophylls need only be included in the diet for the last four weeks prior to the birds' being marketed.

Table 14 Nutritional food additives commonly used for broilers and layers (*Feedstuffs*, 27 January 1975)

Additive	*Uses*	*Pre-slaughter withdrawal time*
Arsanilic acid or Sodium arsanilate	Bloom and feathering	5 days
	Egg production	5 days
	Growth promotion	5 days
	Pigmentation	
Bacitracin	Egg hatchability	None
	Egg production	None
	Growth promotion	None
	Appetizer	None
Bacitracin methylene disalicylate	Egg hatchability	None
	Egg production	None
	Growth promotion	None
Bacitracin, zinc	Egg hatchability	None
	Egg production	None
	Growth promotion	None
	Appetizer	None
Chlortetracyline	Egg hatchability	None
	Egg production	None
	Growth promotion	None
Erythromycin	Egg production	None
	Growth promotion	None
Furazolidone	Growth promotion	5 days
Lincomycin	Growth promotion	None
Nux vomica extract	Appetizer	None
Nux vomica, gentian, anise oil	Appetizer	None
Oleandomycin	Growth promotion	None
Oxytetracyline	Egg hatchability	5 days
	Egg production	5 days
	Eggshell texture and quality	5 days
	Growth promotion	5 days
Penicillin	Growth promotion	None
Roxarsone	Egg production	5 days
	Growth promotion	5 days
	Pigmentation	5 days
Thyroprotein	Egg production	None
	Eggshell texture and quality	None
	Growth promotion	None
Tylosan	Growth promotion	None

CHAPTER SEVEN

Milling and mixing food

Introduction

The earlier chapters have followed through the analogy that the productive chicken or turkey is in some real sense a factory for the conversion of feedstuffs into eggs and poultry meat. In so doing the ingredients of the ration that the poultryman offers his birds have been likened to the fuel that drives the machines on the factory floor and the raw materials which the machines will mould into the desired product.

Whilst the analogy is colourful and, perhaps, helpful, it is still limited in its application. One point that must be crystal clear by now is that poultry food is nearly always a mixture – the fuel and the raw materials do not come separately to the 'factory', but altogether in a single package. A complete poultry ration is delivered to the poultry house and has only then to be put into the feeding troughs in front of the birds for the production lines to start.

This process of preparation of the poultry ration is itself a complex one and is basically common to the rations for all classes of stock. It is, therefore, treated next in the sequence before consideration is given to the design of the different rations for different types of bird. It deals with information frequently omitted from a nutrition text-book which tends to limit itself to balancing the equation between production requirement and nutrient availability, as though regarding the physical preparation of the answer as the concern of someone else quite apart.

In many practical situations today, however, either in sophisticated poultry production schemes where integration of all steps in the sequence is already undertaken, or in developing industries where unified control might also be an advantage, it seems sensible to set down an account of the ways in which the ration can physically be prepared for the birds' use. This is what this chapter attempts to do – relating the milling and mixing of the rations to the whole production organization.

The decision to install a mill or not should be taken following a careful appraisal of the resources available. In addition to the potential savings in production costs, the very real advantage of being able to control the quality of the foods by growing the ingredients locally should be considered.

What level of production justifies the establishment of a feed mill to supply the needs of a specific farm? The answer to this question is not simple. Home, even hand, mixing of feed may be undertaken for a very small number of birds. However, it is often not practical to buy comparatively small quantities of ingredients at an economical price.

Small mill and mix units are available with a bin capacity of from ½ – 1 tonne. Such units are often designed for the production of cattle and sheep feedstuff rather than the highly specialized poultry rations – they are based on the grinding of mixed cereal and on preparing the ration in a vertical mixer. They do not allow the same degree of control over the quality of the ration as does the batch system using a horizontal mixer.

A simple small-scale labour-intensive plant, with a capacity of 1 to 2 tonnes per hour, may be based on a 7½ horse power grinder together with a 500 kg (½ tonne) horizontal mixer. Such a unit could for example be used two or three days per week allowing the mill workers to deliver the mixed rations to the production sites on the other days, and producing 20 to 30 tonnes of food per week, sufficient to feed a laying flock of 40,000 to 50,000 birds. On the same basis a mill with a capacity of 5 tonnes per hour could supply a flock of 200,000 laying birds or 50,000 broilers per week.

A fully integrated broiler group with an output of 5,000,000 broilers per year would require a specialized mill with an output of 10 tonnes per hour. On the other hand an egg unit with a flock of 1,000,000 laying birds would require a mill producing 20 tonnes per hour. (See also pages 109–110.)

The establishment of poultry food mills

MILL DESIGN AND PLANNING

For any food mill, careful planning at the design stage is essential and a comprehensive 'impact study' should be undertaken. The mill, whether large or small, should be designed and sited with care to ensure that it will meet the expected requirements and that it is placed in the most suitable situation for the purpose.

The 'impact study' should consider the effect that the establishment of the mill will have on the area where it is to be sited. The study should also consider the function of the mill, the different systems of milling that are available and also the wide range of milling and mixing machinery that could be used. In addition, the needs for staff accommodation, offices, canteens, control and computer laboratories, and space for parking cars and manoeuvring bulk vehicles, must also be considered.

All these points, and others of relevance to the design and planning of a mill, are now discussed in detail.

FUNCTION

The function of the mill is the production of complete food mixtures or compounds that supply the nutritional needs of the stock concerned.

Each ration should be designed to supply all the known nutritional requirements of the stock for which it is formulated. Usually this will be done on a 'least cost' basis, i.e. the least cost of ingredients per unit weight of ration produced up to specification.

To assist in the process, various supporting services are required, including those of nutritionists to prepare suitable formulations, of bio-chemists to analyse the raw materials and the final food mixtures and of accounting and administrative staff.

Large organizations normally provide all supporting services from their own resources. Farmers' co-operatives and owners of smaller mills usually obtain their scientific know-how from a consultantcy service and this may be available either from the feedstuff suppliers or through an agricultural college or university department.

The actual business of a mill, once the nutritional knowledge has been supplied, is the task of compounding or mixing accurately the formulae provided. It must be emphasized that the grinding of ingredients and the mixing and bagging of the rations is a routine task involving the use and maintenance of mechanical, engineering and electrical machinery which in large mills is both expensive and sophisticated. There is every reason, therefore, to plan thoroughly for its effective use and regular maintenance.

CAPACITY

The capacity of a mill is usually expressed in terms of the output of the mill in mixes or tonnes per hour.

The size of the plant will depend upon the calculated requirements of the numbers of poultry to be supplied and this should be carefully established at the outset. The design of the plant should allow the greatest flexibility in this respect. It is wise to plan for possible expansion and to look five to ten years ahead.

Good-quality milling and mixing equipment, properly installed and maintained, can be run automatically and almost continuously for 20 to 24 hours daily, for up to 260 days per year. Since each 'mix' takes a matter of only minutes, the potential output of the unit can soon be calculated.

At the planning stage, however, it is best to allow some margin of spare capacity, and plans, therefore, may be based on running the mill for sixteen hours a day for 250 days per year. On this basis the mill would be worked for two shifts per day for five days per week. Such a method of calculation allows a built-in reserve, in the form of an extra shift per day or an increased number of working days per week or month.

STORAGE

Storage facilities are expensive and not directly productive so the capital invested in them has to be allocated carefully.

Storage facilities for a month's supply of raw materials on the site would generally be sufficient to allow the mill to continue working even if supplies are temporarily halted due to unforeseen circumstances. Many mills operate on fourteen days' supply of ingredients but the final decision must depend upon the reliability of the mill's sources of materials and the ease with which they can be delivered to the mill.

In those countries where the bulk of the food has to be imported, particularly where there may be shipping problems, three to six months' supply of food ingredients should be readily available either at or near the mill.

The major ingredients will be stored either in bulk silos (see Chapter Seven, pages 117–121) or in the bag warehouse; every effort should be made to receive and store ingredients in bulk, since bulk consignments will be cheaper in price and easier to handle.

Many of the imported raw materials, particularly protein-rich foods and concentrates, arrive in bags. The relationship between the capacity of the bulk silos and the bag warehouse requires careful calculation. These constructions should be sited as near the grinding section of the mill as is convenient. The walls of the warehouses should have sufficient height to allow for stacking and be solid enough to withstand the pressure of bags stacked against them.

The bag warehouse for finished products may be situated underneath the raw-material bag warehouse and this should allow for handling the bags on pallets. Sufficient capacity for three to four days of production will generally be enough if the running of the mill is carefully matched to demand.

The space required for storage in silos may be calculated on the density of the ingredients in terms of weight per unit volume. Cereals are considered to weigh 0·77 tonnes per cubic metre (48 lb per cubic foot) and meals 0·55 tonnes per cubic metre (35 lb per cubic foot).

Space must be allowed in the bag warehouse to stack different feedstuffs separately. Bagged food is best stored on pallets stacked neatly up to four high by means of a fork-lift truck. For this purpose the walls of the warehouse should be 4 to 5 m (16 ft) high.

A useful rule of thumb for calculating the capacity of a warehouse, including space for passageways, turning space for the fork-lift trucks and assuming pallets stacked up to four high, is one tonne of food per square metre of floor space.

The mill design must also make ample provision for the storage of micro-ingredients, and in hot climates this store should be fitted with suitable cooling units.

SYSTEMS OF MILLING

There are two main systems of grinding and mixing ingredients for the production of poultry rations; these are the 'batch' system and the 'continuous flow' or 'volumetric' system.

The batch system

The batch system, under which a complete batch of feed is made with every mix (i.e. one or two tonnes according to the size of the mixer), is the system of choice for the production of high-quality foods. This method is the most satisfactory in respect of accuracy of weighing and mixing. Accurate mixing is essential when it is necessary to incorporate very small quantities of additives into the food mixture.

Another advantage is that the individual ingredients may be ground separately before mixing and in this way the best texture of mash is obtained. Further, the system provides that a rapid change may be made from producing a mix of one formula to one of another; it has genuine flexibility.

Its main disadvantage is the high total cost of installation when compared with the less satisfactory volumetric system.

Volumetric system

The continuous flow or volumetric system of grinding and mixing has the advantage of an apparently low initial cost which is due to the lower number of bins required and to the absence of an accurate weighing system.

The system is based on the supply of raw materials metered directly to the grinding and mixing system on a volumetric basis rather than after the weighing of each individual ingredient. The method is suitable when long runs of non-specialized rations have to be prepared. As a general rule the raw materials are ground together following blending rather than before as with the batch system.

The disadvantages of the volumetric system include the time taken in changing from one formula to another, the fact that the plant is not suitable for producing batches of foods, and is also not really suitable for the production of foods with a high fat content. The system is not very flexible and is more liable to human errors than the batch method. The operator may overlook changes in the density of the ingredients or the fact that one of the ingredients has run out. In this way the formulation of the feed may be unbalanced, with unfortunate results to the stock concerned. A common cause of variation in the volumetric system is the holding-up of an ingredient in the storage bin due to 'bridging' or failure of the material to run under gravity.

Another problem in practice is that adjustments to the rate of flow of ingredients are frequently not carried out immediately and the omission may go undetected for a long time.

BUILDINGS AND LAY-OUT

Careful design of the mill and the lay-out of the various sections are essential to allow a high output at the least cost per tonne of food produced. It is wise to seek specialist advice on the point whether the mill is to be a simple mixing unit or a large-scale plant. In this way advantage can be taken of the latest developments. The advice will cover suggested lay-out plans and flow diagrams of operation to enable further discussions to take place, matching design directly to requirement.

The housing of a small mixing unit may not be a serious problem and it may be possible to find an existing building that will be suitable.

In the case of a large-scale plant with a capacity of 10 tonnes or more per hour, the choice will be between a multi-storey or single-storey building. The multi-storey building is usually the more convenient for it allows the maximum use to be made of the space that must be provided to accommodate the considerable height of much of the equipment to be installed. The different floors can be used for storage of ingredients as well as for offices, toilets, recreation rooms and canteens. Thus the bag warehouse can be situated on the second floor of the building and the warehouse for the finished products on the floor immediately below.

The position of the weighbridge and the weighbridge office requires careful consideration and it is often convenient to site the former as near to the bulk silos and the entrance to the site as possible. The weighbridge should be long enough to accommodate the modern bulk vehicles in which the ingredients may arrive at the mill. In large mills a weighbridge to bear loads of 50 tonnes and of 15 m (50 ft) in length may be suitable.

An important consideration when planning the lay-out of the mill is future expansion. In the case of a mill with a planned capacity of from 1 to 5 tonnes per hour the linear design should be considered. In this arrangement the mill commences with the intake silos and the bag warehouse, followed by the grinding, blending, mixing and outloading sections, in one straight line. This arrangement has the advantage that a second line can be installed alongside at a later date, with the minimum of inconvenience. With two production lines available it is possible to service one or the other without bringing the whole mill to a standstill.

The mill buildings will, for preference, be steel-framed, with the walls made from bricks, concrete, or whichever material will give the necessary insulation at the lowest price. Steel-framed buildings are usually erected in less time than those with brick or concrete supports.

It may well be important to ensure that the mill does not become too hot in the summer and, therefore, the insulation qualities of the building materials are particularly important.

The contractor should be required to specify that the building will be completed in accordance with internationally accepted standards, specifications and workmanship.

Whenever possible the modern mill should make the fullest use of bulk methods of delivery and storage, i.e. silos. The design of the mill should be such that the bulk lorries can be unloaded under cover. The bulk unloading bays should be wide enough and of sufficient height to allow the largest lorries of, say, 20-tonne capacity to be unloaded.

SITING THE MILL

The term 'mill' is used to refer to the entire complex and not merely to that part of the equipment that actually grinds or mills the cereals.

The mill, whether large or small, should be sited to ensure that it is in the most convenient place, as near to existing services as possible in order to minimize the costs of constructing approach roads and of the installation of electricity and water supplies. Generally it is considered to be cheaper to site the mill near to the point of consumption, on the basis that the cost of delivery of compounded foods is usually more expensive than the haulage of the raw materials. Another reason is that the shorter the distance from the mill to the farm the easier it is to maintain the quality of the food (i.e. the mixture is less likely to 'shake-out').

Sites with a high water table should be avoided. If a site liable to flood has to be used then the intake hoppers should be positioned well above the ground, higher than the possible flood level. Placing intake hoppers above flood level will generally be cheaper than trying to provide waterproof hoppers below ground.

At the time the site is being selected the provision of car parks and parking facilities for the delivery lorries should also be considered. The site should be well away from dwelling houses so that mill and traffic noise will not create a nuisance.

STAFF REQUIREMENTS

Accommodation

The space and facilities provided for the staff will depend upon the numbers employed and upon the legal regulations governing safety and recreation. Toilet and canteen facilities are obvious examples.

In a large mill it is an advantage for the mill manager and the chief maintenance engineer to live at, or near, the mill.

The success of any mill, regardless of its size, depends upon the expertise of the person or persons carrying out twelve 'key' operations. The total requirements of a mill with an output of 20

Mill Manager

- Assistant Mill Manager Operations
- Assistant Mill Manager Maintenance
- Assistant Mill Manager Stores
- Assistant Mill Manager Control

Assistant Mill Manager Operations

Responsibilities	Staff
Intake	1
Bag handling	3
Fork lift	2
Grinding	1
Blending	1K
Pelleting	1K
Packing	1
Finished products	2
Fork lift	2
Outloading	1

Assistant Mill Manager Maintenance

Responsibilities	Staff
Electrical	1K
Mechanical	1K
Vehicle	1K
Eng. stores	1
Boiler	1
Drivers	

Assistant Mill Manager Stores

Responsibilities	Staff
Raw materials	
Deliveries	
Typist	1
Clerk	1K

Assistant Mill Manager Control

Responsibilities	Staff
Sampling	
Punch cards	
Micro-ingredient stores	

Total staff	
Managerial	5
Mill workers	22
Total	27 + drivers

K = staff requiring special training

The above figures are based on one eight-hour shift daily. The total staff to be increased by sixteen whenever the mill works two shifts, i.e. those under the operations manager plus one maintenance engineer.

Fig. 5 Staff and jobs involved in a 20-tonne per hour feed mill. (Based on European standards for well-trained and experienced staff)

tonnes per hour are outlined in Fig. 5 where the 'key' workers have been indicated by the letter 'K'.

Training

A continuous programme of training should be established within each mill and all 'key' personnel should have deputies already trained to carry out their functions in the mill. The mill workers too should be trained in as many routine tasks as possible to ensure their interchangeability – e.g. the blender and the pelleter should be equally proficient at each other's job.

The aim of the training programme should be to provide a constant source of trained and experienced staff and to allow the opportunity of promotion from within the company, for those workers with the initiative and ability to work hard and efficiently. It should be possible for one of the 'key' workers to become the manager of the mill. In the same way a ladder of promotion is offered to labourers who aim to become one of the 'key' workers.

SAFETY PRECAUTIONS

A good deal of attention must be given to safety precautions in the mill and the following features should be incorporated in the original specifications.

Secure covers should be placed on all machinery drives, and safety grids should be installed immediately underneath all manhole covers in the tops of the bins and silos. All open spaces should be protected by handrails, and all holes in the floors around the elevators, spouts and machines should be filled.

Electrical stop buttons should be placed at suitable points on all the elevators and back-stops fitted to prevent the elevators from reversing.

Dust-explosion protection should be fitted to the exhausts of all grinders and an efficient system of dust removal installed. The fire precautions should include the provision of fire extinguishers and fire escapes, additional to the normal stairs and exits.

FLOW DIAGRAMS

The task of planning a new mill is made much easier by the

preparation of a flow diagram, sometimes called a flow sheet or flow plan, to illustrate the route taken by the ingredients from their arrival at the mill to their exit as balanced foods. The flow diagram also specifies the major items of equipment and machinery that go to make up the mill.

Alternative flow diagrams should be prepared before the final plan is agreed.

ACCEPTANCE TEST

A comprehensive acceptance test should be included in the contract for the erection of a mill, large or small. In this way the purchaser can be satisfied that the mill is capable of being run at the specified rate.

A suitable acceptance test for a large plant is to run the mill for a continuous period of twenty-four hours under the supervision of the contractor. The agreement covering the test should include a clause that any stoppages during the test run that are due to the fault of the contractor will be considered as part of the test period. On the other hand the time taken by stoppages due to the buyer's fault should be added to the test period. In the event of stoppages due to the contractor exceeding four hours, or if the throughput fails to reach the required standard, the contractor can be given a period, e.g. thirty days, during which to make any changes or replacement of equipment prior to re-running the test.

Milling and mixing machinery

INTAKE SILOS

The silos in which the bulk ingredients, generally the cereals, are stored at the mill, should be of sufficient size to allow the storage of at least two but preferably four weeks' supply (see Chapter Seven, pages 109–111).

The precise requirements concerning the capacity of silos should be calculated before they are purchased by relating the size to the quantity and density of the materials to be stored. A silo that will hold 500 tonnes of wheat will hold a much lower quantity of poor-quality barley which has a higher volume per unit weight.

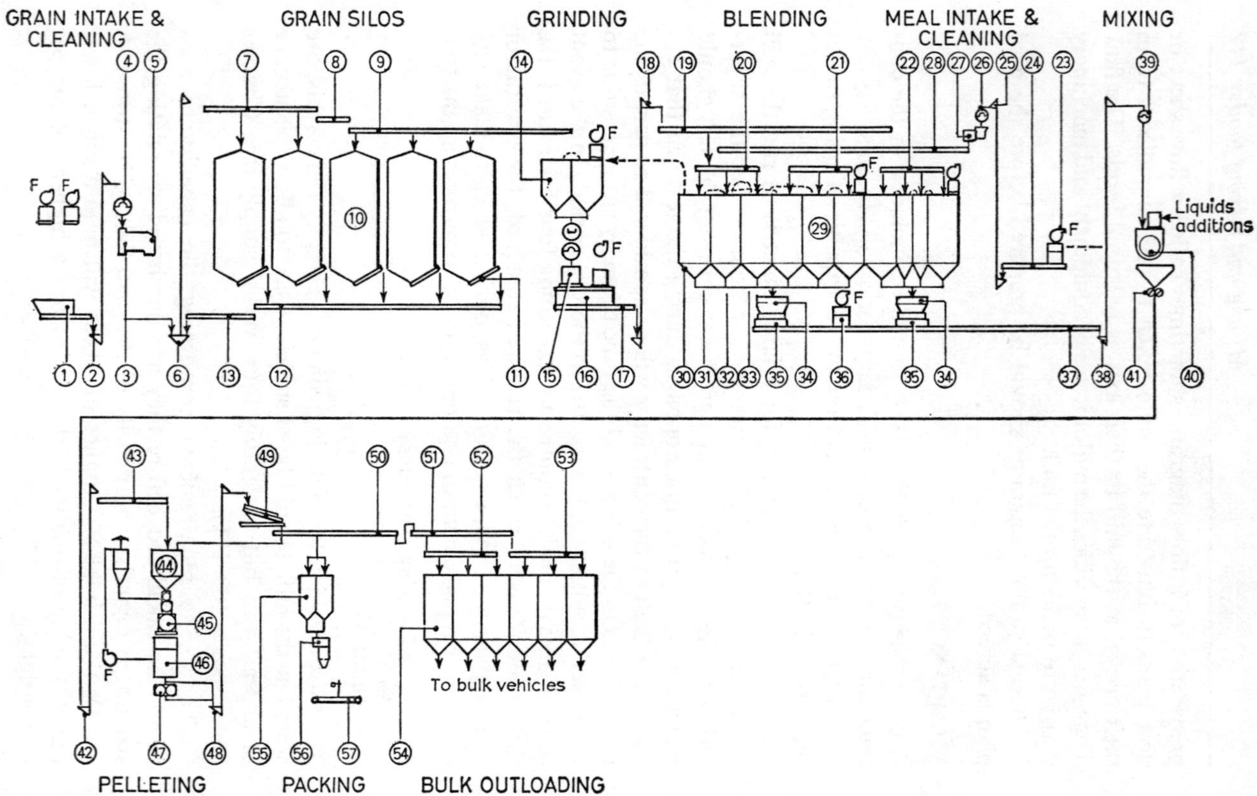
GRAIN INTAKE & CLEANING
GRAIN SILOS
GRINDING
BLENDING
MEAL INTAKE & CLEANING
MIXING
Liquids additions
To bulk vehicles
PELLETING
PACKING
BULK OUTLOADING

Fig. 6 Flow diagram for a 10-tonne per hour food mill

1 Heavy duty intake hopper with safety grid
2 Conveyor
3 Elevator
4 Cleaning magnet
5 Rubble separator
6 Elevator
7, 8 and 9 Conveyors
10 Storage soils
11 Discharge augers
12 and 13 Conveyors
14 Grinder service bins
15 Hammer mill
16 Grinder under hopper
17 Grinder discharge conveyor
18 Elevator
19, 20, 21 and 22 Conveyors
23 Dust control unit
24 Conveyor
25 Elevator
26 Magnet
27 Sifting machine
28 Conveyor
29 Blending bins
30, 31, 32 and 33 Bin discharge
34 Hopper weigher
35 Weighing under-hoppers
36 Additive tipping unit
37 Conveyors
38 Elevator
39 Magnet
40 2,000-kg mixer
41 Bin agitator
42 Elevator
43 Conveyor
44 Pellet service bin
45 Pellet mill
46 Pellet cooler
47 Pellet crumbler
48 Elevator
49 Sieve unit
50, 51, 52 and 53 Conveyors
54 Bulk outloading bins
55 Bag packing bins
56 High-speed packing unit
57 Automatic bag stitcher

The capacity of the silos should be stated by the vendor in terms of their cubic capacity or in tonnes of the materials to be stored. In this way the capacity of the silos can be checked by the buyer. A common means of expressing the required silo capacity is: each silo to hold 500 tonnes of cereals of 0·77 tonnes per cubic metre (48 lb per cubic ft), for cereals with a moisture content not exceeding 14 per cent.

The working capacity of a silo usually refers to the quantity of food the silo will hold when full and level. For wheat (at 0·77 tonnes per cubic metre) the working capacity will usually be about 8 per cent less than the 'coned' capacity.

Silos vary both in design and materials of construction. The most popular silos for general use are round and these are available with up to 2,500 tonnes capacity. Rectangular silos are particularly useful where large quantities of cereals have to be stored. Such silos are often erected in the form of an independent building and may be divided into a number of compartments and equipped with drying facilities. In all large silos it should be possible to recycle the cereals, i.e. move the grains from one silo to another by means of conveyors. This arrangement is particularly valuable when dealing with cereals that may have an above-average moisture content.

The type and quality of the materials used in the silos should always be stated and be up to an acceptable international standard. Most silos are available in a pre-fabricated form. Corrugated iron is a very popular material. Panel silos, as the name implies, are made up of strong galvanized iron, wooden or concrete panels, that are fitted together and are suitable for silos of almost any capacity. Collapsible plastic or canvas silos are also available and are noted for their portability and ease of erection (see Plate 7).

The inside walls of the silos should be smooth and suitable for any of the ingredients that may be used in the mill.

Bin activators or vibrators may be necessary to simplify the discharge of stored material. They overcome the problem of bridging, and are particularly useful in situations where wide, low and rectangular silos have to be used.

A contract for the installation of silos should ensure that they will be complete with all the necessary augers, conveyors, belts and elevators to allow the silos to be filled or emptied at a specified rate. A suitable rate of intake by the silos for a commercial mill is 60 to 100 tonnes per hour. Such a speed will allow bulk delivery vehicles to be unloaded rapidly and thus facilitate a quick turn-round of lorries. To accomplish this, the heavy-duty intake hopper should have a capacity of at least two tonnes. A strong grid should cover the intake hopper as a protection from damage by vehicles. The silos should also be equipped with the necessary walkways which may also be used to provide support for the conveyors.

The intake hoppers of the silos are usually situated entirely or partially below ground level to a depth of 3 m and must, therefore, be waterproofed and protected against surface water. It is also wise to protect the intake hopper from the effects of weather and this can be done by erecting a simple brick or steel-framed building. The building can also serve as an office for the person in charge of the silos and possibly also for the weighbridge attendant. In hot climates the building should be well insulated.

The system of delivery and discharge of materials should also include a means of rapidly emptying and recharging the silos to give a complete turn-round of the ingredients. This is a valuable asset as a means of controlling the problem of overheating which may occur if the grain has too high a moisture content.

In planning a new mill the utmost use should be made of automatic means of conveying both the ingredients and finished foods. Thus the reception silos should be equipped with conveyors to supply the grinding bins of the mill, first taking the ingredients over the scales and cleaning units on the way.

CLEANING EQUIPMENT

Cleaning or screening of ingredients involves the separation from the food materials of all foreign matter which may include stones, bits of metal, straw, string and pieces of wood, all of which can cause considerable damage to the grinding and mixing machinery. This operation also involves the removal from the ingredients of weed seeds and other contaminants, e.g. corn-cob meal from maize. This is accomplished by separators, and both sieve and magnet types are available.

The sieves screen the ingredients on the basis of size of particles, the small grain particles passing through the apertures while the remainder is retained. Sieve separators are installed at the intake to the bulk silos to enable the cleaning process to commence immediately the materials reach the mill.

Weed-seed separators are particularly necessary when specialist foods are produced, to avoid the presence in the final mixture of any seeds that may have injurious effects on the stock.

A separator to remove the rubble should also be installed to sift the ingredients before they reach the grinding system of the mill.

Magnetic separators remove pieces of iron which have found their way into the ingredients. Types available include drum, band and spout magnets. These magnets should be placed wherever there are entry points into the conveying or mixing systems.

Brush separators are usually placed in the system beyond the grinding process; they make use of a high-speed sifter which brushes the feed through a sieve. That portion which is left above the sieve is rejected.

GRINDING

The grinders used for breaking down the cereal grains are often called 'hammer' mills because of the 'beaters' or 'hammers' that they contain. These hammers play an important role in reducing the particle size of the grains and are liable to considerable wear.

The operation of grinding involves a number of different processes: impact, cutting, crushing and attrition. Attrition refers to the process in which the particles are reduced in size, by friction against each other and against the parts of the grinder itself, before they pass through the exit screens. The final particle size is largely determined by the size of screen used. Worn screens soon lead to variations in particle size or 'grist'. Ample supplies of spare hammers and screens must always be held in stock.

Screens are available either in metric or English sizes and both are shown in Table 15. Fine grinding results from the use of screens up to 2 mm and coarse grinding from screens of from 2 to 5 mm.

The output of the grinder depends upon the type and quality of the raw materials, upon the design and power of the mill, and upon the size of the perforations of the grinding screens. Materials vary considerably in the time they require for grinding. That, for example, of barley being about twice the time required for maize.

Table 15 Screen sizes, with approximate equivalents

English	*Metric*
1/20 in = 1·25 mm	2·50 mm = 1/10 in
1/16 in = 1·6 mm	3·00 mm = 1/8 in
1/8 in = 3·17 mm	4·00 mm = 3/20 in
1/4 in = 6·35 mm	5·00 mm = 1/5 in
	6·00 mm = 1/4 in

Large grinders are far more efficient than small models. Thus a farm machine of 5 hp will grind about 200 to 300 kg of maize per hour through a 3·17-mm screen (needing about 20 hp per tonne) whereas a 40-hp model with a suction system will grind about 5 tonnes of maize through a 3·0-mm screen which is equal to 8 hp per tonne. A modern 75-hp grinder will process up to 10 tonnes per hour through a 3·0-mm screen.

The machines chosen must be capable of maintaining the full capacity of the mill. To allow the mill to be as flexible as possible it is generally considered best to install at least two grinders, so that production is never brought to a halt by mechanical failure at this stage of the operation.

In a mill with a capacity of 20 tonnes per hour, each of the two grinders should have a throughput rate of 15 tonnes per hour of maize with a normal moisture content. The whole operation of grinding may be automated so that the machines operate for long periods, perhaps overnight, without attention.

Mixtures of dust and air in a grinding mill can, under certain circumstances, be explosive. To overcome this danger the exhausts of all grinders should include dust-explosion protection and an effective method of dust control should be installed in the mill.

The holding bins should be placed to supply the grinders automatically and, in the case of the 20-tonne-an-hour mill, eight bins each of 30 tonnes capacity will normally be satisfactory. For a plant of 1 tonne per hour two supply bins of 10 tonnes capacity will be needed.

In the process of grinding some loss of weight of the materials takes place in the form of dust and by evaporation. The loss due to evaporation may amount to 3 per cent of the throughput. Humidifying machines are available which can restore the evaporative loss and also reduce the amount of dust in the atmosphere.

THE BLENDING SYSTEM

The bins and equipment involved in the supply of the ingredients to the mixers, the mixing machinery itself and the conveyors and bins for the discharge from the mixers, are collectively referred to as the blending system.

In a large mill the whole blending process is best arranged entirely automatically. In smaller mills mixing 1 to 5 tonnes per hour some parts of the system may be operated by hand.

A number of different automatic systems are available. In the simplest form, the operator can set the requirements for the mix on a control panel and, by pressing control buttons, arrange for the ingredients to be delivered to the mixer.

The automation of the blending system can be taken a step further by the use of punched cards. In this case the cards are punched with the formula requirements and fed into a panel reader; from this stage onwards, the ingredients are supplied, in succession, to the mixers after first passing over the weighing scales.

In very large mills, the process can be carried out completely automatically with all the operations controlled by a computer.

The blending bins

The blending system commences with the blending bins and includes the conveyors that carry the ingredients to the mixers, the scales and all the other items necessary for accurate working.

The blending bins of the batch-mixing system allow a much greater degree of accuracy in the mixing of rations and gives a degree of flexibility which the continuous-flow system (not using blending bins) does not have. The bins should be of sufficient size to allow grinding of cereals to continue overnight when the mixing plant may not be working.

The numbers and size of the blending bins depends upon the level of use of the ingredients and upon the output of the mill and must at least cover the normal daily output. A simple 1-to-2-tonnes-per-hour mill may, for example, have six bins each with a capacity of 5 tonnes. In a mill with an output of 10 tonnes per hour ten bins, each of 15 tonnes capacity, would be adequate. A mill with an output of 20 tonnes per hour could be provided with twenty bins. A convenient arrangement is eight bins each of 30 tonnes for the cereals, eight bins each of 15 tonnes for the protein-rich ingredients, and four bins each of 10 tonnes for supplying the pre-mixes and concentrates.

Mixing

The accurate mixing or blending of the raw materials into high-quality poultry foods requires specialized machinery that carries with it a guarantee of thoroughness of mixing. Accuracy of mixing is particularly important for the micro-ingredients that have to be included in the mixture, often in amounts of grams per tonne. A suitable degree of distribution is 125 parts per million.

Accurate weighing is essential to ensure that the formula is adhered to and to allow subsequent checks to be made on the efficiency of the ration.

The accuracy of the mixing should be checked as part of a regular quality-control programme. Undermixing and overmixing can both lead to uneven distribution of the ingredients. Over-mixing can result in segregation or 'de-mixing' of the food particles.

Mixers range in size from those capable of 0·5 tonnes per batch to the commercial mixer of 2 tonnes.

The three types of mixers are the horizontal, the vertical and the continuous mixers. The mixer of choice for the production of specialist rations is the horizontal type which provides a much higher standard of mixing than the vertical machine and also has a higher production rate.

High-capacity horizontal mixing plants, with a capacity of 1 to 2 tonnes per mix, may complete the whole process of supply, mixing and discharge in three to five minutes. Horizontal mixers are also suitable for mixing fats and molasses. They are simple to use as part of a three-tier system with the supply bins above and a weighing unit immediately beneath.

Vertical mixers generally take from fifteen to twenty minutes to complete the mixing process. These mixers are not really suitable for use with fats and molasses which tend to adhere to the sides and thus reduce efficiency. In addition, vertical mixers are not considered satisfactory for the inclusion of micro-ingredients where very small quantities have to be mixed in the ration.

Continuous mixers may be used in a volumetric system. Such mixers are designed to receive ingredients that have already been blended and are not suitable for use with 'straight' ingredients.

Farm mixing

Farm mixing of rations can be a most efficient means of using home-grown supplies of feedstuffs and of keeping the cost of feeding to the minimum on farms where the total consumption of foods reaches, for example, 10 to 20 tonnes a week. Farmers following this system may well purchase a protein concentrate and so will have only, say, half a dozen ingredients to consider.

In its simplest form a farm unit can consist of a small grinding machine and a mixer. Combined mill and mixing units are now available that have been specially designed for the farm mixer, and these units are comparatively inexpensive.

A simple unit of the type outlined would rely on farm labour for the supply of the raw materials and delivery of the finished food but, in many cases, the unit could be improved at little extra cost by the installation of meal-holding bins.

On some farms special hoppers have been designed that can receive the food direct from the mixer and are then moved to the house in which the food is to be used. Such semi-bulk bins are generally moved by a fork-lift; these bins may hold half a tonne of meal.

Equipment to allow the food to be carried automatically is now available in many forms. On some compact, small, intensive farms equipment has been designed to allow the food to be taken direct from the meal-holding bins to the automatic feeding system installed in the poultry houses.

MIXING FATS AND MOLASSES IN THE RATION

Fats

Animal fats and vegetable oils are frequently included in poultry food mixtures at levels of 1 to 3 per cent. Special equipment allows a further 3 per cent to be sprayed on to the pellets but results in the production of pellets that are rather soft.

Fats may be delivered in drums of about 200 litres (40 to 50 gal.) in a solid state. A gallon of fat weighs about 3·4 kg (7·5 lb). The fats must be melted before use and this may be done by inserting an immersion heater into the drums or by the use of other specialized equipment.

In large mills, bulk tanks for the storage of fats are desirable, e.g. two tanks each of 400 to 500 litres capacity. The fat may be kept in liquid form between 40° and 50° C (100° to 120° F), at which temperature it is almost ready for immediate use.

Fats may also be delivered in specially designed bulk transporters in which they are kept in a liquid state. In this way the fat can be pumped directly into the bulk tanks at the mill.

Liquid fat at a temperature of 70° to 80° C (160° to 170° F) is sprayed directly into the meal at the top of the mixer. Specially designed sprayers are available for this purpose. Only good-quality fats should be used. Those with insoluble products are liable to cause blockages, particularly of the spray nozzles.

An emergency electricity supply should be available to help to keep the fat liquid during a power breakdown.

Molasses

Molasses may be delivered to the mill either in drums or by bulk transporters and is commonly included in mixes at the rate of 2·5 per cent of the total mixture. Specially constructed, smooth-sided concrete bins or steel tanks are used for storage. The capacity of the tanks is related to the rate of use of the molasses but should be such that a bulk transporter can unload quickly; e.g. if the molasses is delivered by 20-tonne tankers then the mill will require at least two tanks, each of which should have a capacity of a minimum of 30 tonnes.

Molasses, like fat, has to be heated before being incorporated in the food mixture. The pumping temperature required is 40° to 50° C (100° to 120° F) at the point of incorporation into the food.

The boiler

An efficient means of producing high pressure steam is required in all mills and is particularly necessary for the mixing of molasses and for pelleting. The equipment should include suitable water softeners so that the boiler does not become coated with a calcium 'scale' on the inside.

A manufacturer's certificate should be obtained stating that the boiler has been tested to twice its normal working capacity and found to be satisfactory.

The compressors

Compressed air will be required to operate many parts of the milling and mixing machinery and, possibly, for conveying foods pneumatically. The compressors are usually housed in a separate room or building together with the boilers.

CRUMBS AND PELLETS

The meal emerging from the mixer may have to go through one further process before it reaches the output unit; it may be passed through a pelleting unit to be turned into either pellets or crumbs which are often favoured at the poultry farm because of the ease of handling them, the reduction in dust and the increased intake of pellets over mash by the chickens.

Pellets should be firm and tough so that they will withstand handling in bulk and not become broken down by the automatic systems of feeding employed on many farms.

The most suitable grist of meals to be used for pelleting is classed as medium to fine, i.e. that passing through a screen of 2·5 to 3·0 mm. Pellets made from coarse-ground meals tend to break easily (see Chapter Seven, pages 122–124).

The production of pellets and crumbs consists of four different processes. First, the meals have to be prepared for pelleting by a procedure known as steam conditioning which softens them. The food is heated to 60° to 80° C (140° to 180° F) with a moisture content of about 17 per cent. The length of time for which the materials are held at the highest temperature is relatively short so that there can be only marginal impact on the quality of the food caused by any 'cooking' effect. In fact, claims are made that the steam treatment changes the starches into sugars and thus makes them more readily available to the stock. However, gelatinization of the starches requires a temperature of 82° C, and at this temperature vitamin destruction may well occur (see pages 61–62). Too much steaming also causes excess moisture to be absorbed into the meal and results in soft pellets.

The second stage is the pressing of the meal through a mould to turn it into pellets. The rate of this process depends upon the power of the plant and the size of the pellets being produced. The standard size of pellets for adult poultry is 4·7 mm ($\frac{3}{16}$ in) diameter. Growers' pellets are usually 3·00 mm ($\frac{1}{8}$ in). The process is very power-consuming and the mixing unit will usually produce meal faster than it can be pelleted. The lay-out used must make it possible to hold the meal in front of the machine or make some proportion of the output bypass it altogether.

The third stage in the production of pellets is cooling; the coolers are best placed immediately underneath the pelleting machines. Cooling is required at once to prevent the pellets from sweating and then becoming mouldy.

The final stage is the production of crumbs and this involves passing the pellets through the crumbling machine. Crumbs are made by fracturing 4·7-mm pellets so that each crumb has a piece of the outside of the pellet to give it rigidity.

Pelleting machinery is available in various sizes; thus a mill with a pelleting plant capable of producing 10 tonnes an hour of 4·7-mm pellets will have a crumbling plant that produces 6 tonnes of chick crumbs an hour. The reduced output of crumbs is due to the screening off of the so-called 'fines' that are returned for re-pelleting.

Regular checks should be made on the quality of the pellets being produced.

THE OUTLOADING SYSTEM

Outloading bins

Meals, pellets and crumbs should flow automatically to the holding bins for finished products. These bins should be situated near to the bulk unloading point and to the packing unit. A suitable arrangement for a mill of 20 tonnes an hour is eight holding bins each with a capacity of 20 tonnes.

Augers and conveyors may be used to allow the finished products to be discharged, after passing over the weighing scales, directly into the bulk delivery vehicles, whenever this is a convenient alternative to bagging.

Packing in bags

Bagging should be arranged automatically and packing, stitching and sealing can be arranged at a rate of up to ten bags a minute, with output monitored by an automatic counting device. Hand-stitching machines are also available.

Meals are generally packed in 25-kg or 50-kg bags. The 25-kg bags are the more convenient to handle on the farm.

An automatic method of sealing (instead of stitching) is available and is known as valve packing. The system is suitable for large plants.

At the time the bags are sealed a label can be included giving details of the contents of the food, the date of mixing and other relevant information. The bags themselves can also have the information printed or stamped on them.

Additional handling equipment

Other individual items will often be required to ensure the smooth running of the plant. In the bag warehouse it may be useful to have a special elevator for stacking sacks and also a horizontal conveyor.

A fork-lift truck is a valuable asset for moving pallets stacked with finished meals and pellets (see Plate 8).

Pneumatic-suction conveying systems can play a useful role in the modern mill. The advantages reported include prevention of the separation of ingredients that sometimes occurs when foods are discharged in bulk to silos by more conventional methods.

Pneumatic conveyors are easy to install and can bypass difficult corners. They are almost self-cleaning and therefore require little maintenance.

WEIGHING

Accuracy of weighing is of particular importance in the mill and the weighing system should carry a guarantee of accuracy. In the case of commercial mills a suitable standard is an accuracy of $\pm$ 0·25 per cent. The scales and the print-out facilities should be in both metric and English weights.

The first item in the comprehensive weighing system of a modern mill will usually be the weighbridge and its best position is near the entrance to the mill. Bulk lorries carrying 20 tonnes or more of materials are much used nowadays and require a weighbridge capable of taking up to 50 tonnes.

The weighbridge gives the gross weight of the load including the weight of any non-edible material. The latter can be high and to check on this it may be necessary to arrange for the materials entering the silos to be weighed before and immediately after cleaning. This is desirable in order to allow the proper control of stocks.

Accurate scales are required in batch-mixing plants to ensure that the correct formulations are being prepared. Very sensitive scales are necessary for weighing the micro-ingredients before pre-mixing.

The weighing system is completed by the automatic weighing and bagging unit or by the bulk outloading weighing unit. These can be installed to give either the gross or the nett weight; the latter are to be preferred, although the former may be more readily available.

In small plants platform scales will be required for weighing the ingredients before mixing and for weighing the finished products.

THE DUST CONTROL SYSTEM

Air filters and dust-exhaust equipment should be installed as necessary to give a dust level of not more than 4 milligrams per cubic metre of air in all the working areas of the mill. At the same time the system should prevent concentrations of dust in the mill from escaping to the outside atmosphere.

The system also should be designed to draw off moist, hot air and thus prevent the build-up of moulds whilst, at the same time, keeping the machines cool. Excessive loss of moisture from the food should also be prevented by an efficient system.

Modern systems of dust control can be most efficient. The old system of cyclones with dusty sleeves has now been replaced by modern suction methods which rely on the high-speed passage of air and thus tend to be self-cleaning.

If possible, the equipment used for dust removal should form an integral part of the machines themselves. Thus modern grinders allow for dust removal as well as suction delivery of the ground ingredients direct from the grinders. In the same way separators and pellet coolers have their own inbuilt dust-removal systems.

Dust control in the mill is much influenced by the cleanliness of the ingredients and by the fineness of grinding. Poultry prefer coarsely ground mashes and this is also helpful in controlling dust.

Limestone flour can be an important source of dust if it has been incorrectly processed. Granular limestone is preferred for use in laying mashes.

The efficiency of the dust-control system should be regularly checked. Collected dust should not be re-cycled for use in food mixtures if there is the least danger that it has become contaminated in any way.

PRE-MIXING

Additives, including vitamins, trace minerals, drugs and antibiotics have to be included in the food mixture in very small amounts, e.g. parts per million. The best way to achieve this is to pre-mix the additives with one or more other ingredients acting as carriers or spreaders.

The pre-mix or concentrate can be made up for inclusion in the main mix at whatever proportion is convenient for the particular plant, e.g. for use at 1, 5 or 10 per cent of the total.

In small mills the pre-mixing can be carried out by hand. In a large modern mill a complete pre-mixing unit may be installed. Such a unit for a mill with a capacity of 10 tonnes per hour may consist of:

(a) A pre-mix supply bin of 10 to 20 tonne capacity for the carrier.

(b) A set of hopper bin scales of 500 kg capacity that receives the ingredients and delivers them to the mixer.

(c) A horizontal mixer of 500 kg capacity and guaranteed accuracy of mixing.

(d) A pre-mix discharge bin that is placed immediately under the mixer, being capable of taking its full capacity.

(e) A 50 kg belt-fed weighing and packing unit which is primed directly from the pre-mix discharge bin. Alternatively, the pre-mix can be delivered direct to one of the blending supply bins.

Maintenance

HYGIENE IN THE MILL

A high standard of cleanliness is required in the mill to prevent it from becoming a breeding ground for bacteria, fungi, insects, rats and mice.

Regular sweeping, removal of the sweepings and good dust-control procedures are necessary in maintaining a clean mill.

Rats, mice, insects and micro-organisms thrive in dirty conditions and, if not controlled, can give rise to contamination of the rations and to outbreaks of disease amongst the stock feeding on them. Salmonellosis is frequently caused by dirty rations.

MACHINERY MAINTENANCE

The process of milling and mixing employs expensive machinery that requires regular and correct servicing. The machines installed will have power units that range from the small ¼-hp motors to large engines of 250 hp or over.

The maintenance engineers and fitters responsible should be trained in maintenance by the suppliers of the equipment that it is their job to service. In the case of new mills and new equipment it is desirable for the maintenance engineers and fitters to assist in the actual erection of the plant. In this way it can be ensured that all are completely familiar with the lay-out and all maintenance requirements. Routine maintenance should be carried out regularly during the time the machines are not in work.

The chief maintenance engineer should carry out regular inspections and tests of all the machines and equipment under his care. To assist him a special 'check list' should be prepared; the list should be designed specifically for that particular mill. It should include a complete flow diagram of the mill and also a note of the position of all the machinery that requires extra attention. The check list should cover the complete production line, the boiler and compressors and the cleaning and dust-control systems. It should also include inspection of all handling equipment, conveyors and lifts and testing for accuracy of all scales.

ELECTRICAL MAINTENANCE

Regular maintenance of all electrical equipment is essential to keep the mill free from breakdowns and is also a vital part of fire precautions.

In the case of new installations a certificate should be obtained giving the calculated electrical-power consumption of the complete plant, itemized by machines. The standard kilowatt cost of the rations mixed can then be established on a per-tonne basis.

The electrical engineer should be present during the installation of new electrical equipment to ensure that he is completely familiar with the installation and the wiring.

Manufacturers' certificates of testing and efficiency should be obtained with all new equipment and, in particular, for control panels, starter motors, generators and other basic equipment.

All wiring should be up to international standards. Some forms of plastic-covered wires may be eaten by rodents, thus creating a risk of fire or breakdown, and should be avoided. In hot climates, all electrical equipment should be tropicalized and protected against dust. A complete wiring plan should be prepared, kept up to date and prominently displayed. There should also, in the larger mills, be a control panel complete with mimic diagrams that give an indication that all circuits are working properly throughout the mill. The details of the power supply should always be clearly stated in all specifications, e.g. single- or three-phase, 240 watts and 50 cycles (see Plate 9).

In modern mills the plant should include a stand-by generator to cover the emergency services including lighting, the fat and molasses plant and, possibly, the cooling of the micro-ingredient store.

An electrical check list should be prepared that includes a list of all the electric motors, electrical control panels and equipment, electro-magnets, switchboards and their position. The check list should also make reference to the checking of all electrical boxes for freedom from dust and missing covers and the checking of the fork-lift trucks, which are likely to be electrically operated.

VEHICLES

The aim must be to keep all vehicles in good running order at all times and this involves regular and correct maintenance. Maintenance should always be carried out during non-working hours in order to create the minimum interference with the normal functioning of the mill.

Vehicles designed for the transport of food in bags may be fitted with a mechanical hoist to facilitate loading, or should be loaded by a fork-lift and pallet system.

A full programme to ensure the regular maintenance of the vehicles should be prepared. Driving large articulated lorries requires training and skill. It is wise to ensure that all drivers attend a driving course. Suitable courses are sometimes organized by the manufacturers.

Bulk delivery vehicles range in size from those holding 2 to 3 tonnes to those with a capacity of 20 tonnes or more. Methods of discharge are by gravity or by pneumatic means.

Collection of grain direct from supply farms can be greatly speeded up by collecting in bulk. Vehicles should carry an intake auger for use on arrival at the farm. In this way the farmer does not have to make any special arrangements for the dispatch of cereals in bulk.

Vehicles specially designed for bulk delivery of molasses are also available. Fat may be delivered in specially designed vehicles in which the fat is heated and kept in a liquid state to enable its immediate discharge on arriving at the mill.

The full value of bulk delivery vehicles will not be obtained unless the farms and mills are also designed for bulk food, though one of the most popular types of delivery vehicle is that which can also be quickly adapted for bag carrying.

Books giving details of vehicle design and maintenance can be obtained from the manufacturers. It is wise to specify that these should be supplied, in the desired language, at the time the vehicles are ordered.

Bulk feed trailers, designed to be towed by a farm tractor, are a useful means of transporting food direct from the farm mill to the bulk bins that supply the individual poultry houses. Trailers with a capacity of up to 8 tonnes may be equipped with an auger along the base of the hopper and a delivery auger of 8 m in length. The mill may be designed to allow batches of food to be discharged directly into the trailers thus eliminating the bulk outloading bins.

Bulk bins or hoppers are usually installed at a convenient point outside the poultry house and equipped to allow the food to be delivered either pneumatically or by auger. The bin should be large enough to hold one week's supply of food for the house. Ten-tonne bins are very popular for use with broiler houses of

20,000 birds although bins of 17-tonne capacity are often preferred. Round, square and polygon hoppers are the most popular and these may be made of wood, galvanized iron or plastic (see Plate 10).

Stock control

INVENTORIES

Inventories should be kept of the quantities and types of all expendable items used in the mill. The daily recording of food ingredients and additives is an essential requirement of any mill. However, up-to-date records are also necessary for stocks of drugs, spares for the milling and mixing machinery, vehicles and stationery. In large establishments the control of stocks of all types can best be accomplished by the use of a computer.

The system should detail the calculated levels of ingredients and compounds held at any time; forecasts of ingredient requirements and capital employed; forecasts of food-mix costs according to raw material prices. In addition the computer system should provide for an automatic print-out and, if required, a video display of these data.

Stock control will be required in four separate sections though, in small mills, these may be the concern of one man. Large plants may require four individuals to be responsible separately for: (a) food ingredients; (b) micro-ingredients, including drugs; (c) general stores; and (d) the administrative stores.

MICRO-INGREDIENTS

Some micro-ingredients may lose their value on storage and for this reason it may be necessary to install cooling facilities in the storage area. When this is done a daily record of the maximum and minimum temperature should be kept.

The store will usually be under the control of the person who is also responsible for the operation of the quality-control procedures of the mill. It should be kept locked to avoid the accidental loss of dangerous drugs. All the micro-ingredients should be distinctly labelled to overcome the possibility of mistakes.

CHEMICALS

The non-food chemicals, e.g. the fungicides, insecticides and the chemicals used for controlling rodents, should be stored well away from the micro-ingredient store so that accidental contamination of food ingredients by noxious substances does not take place. In this regard care is necessary in the disposal of wastes including dust and sweepings.

MACHINERY AND VEHICLE SPARES

Orders for new vehicles or machines, should include a supply of vital spares. If these have to imported, at least one year's and preferably three years' supply may be required even if this ties up a lot of capital.

THE ADMINISTRATIVE STORES

The administrative stores should ensure that up-to-date catalogues are kept for all machines and items used in the mill. When an emergency occurs it is most helpful to have the catalogue numbers of the parts required immediately to hand.

The administrative stores may also be responsible for the provision and maintenance of photo-copying facilities suitable for reproducing pages from books and periodicals, and to provide copies of administrative directives for rapid circulation.

CHAPTER EIGHT

Formulation and feeding programmes

Ration formulation for different classes of poultry stock

GENERAL PRINCIPLES OF FORMULATIONS

The basic problem of nutrition is the solution of an equation which can be simply, perhaps too simply, stated: The ration fed is equal to the determined requirement of the animal for nutrients.

A moment's thought makes it clear that this statement has left many questions unasked on either side of the equation. On the bird's side, for example, it will be necessary to know:

(i) Have the requirements been adequately determined?
(ii) Do they relate to the degree of activity or productivity which is going to be expected of it?
(iii) What amount of this ration will the bird eat in a specific time (usually a period of twenty-four hours)?
(iv) Will this intake provide the requirements tabulated?

In respect of the ration formulation, the following information is required:

(v) Are feedstuffs available to provide the requirements specified for energy, amino acids, vitamins and minerals?
(vi) Can they be mixed together to give a ration of good keeping quality, that can be easily handled and which will be readily eaten by the stock for which it has been made?
(vii) Can it be manufactured at a favourable cost?

The person formulating the ration, whether he is a professional nutritionist working for a feedstuffs company or whether he is a poultry farmer planning to mix a formula based substantially on home-grown and locally available ingredients, has to have all this information. Different people may put different emphasis on the questions set down. The nutritionist will be trying to strike the first balance (between the sides of the equation) in respect of common units of energy or protein. The compounder is likely to be more concerned with the relative costs of different ingredients which might be used in the formulation. The farmer will be concerned to know whether the ration is readily eaten and if it does the job asked of it.

All these interests are involved in the problem of balancing the ration. More will be said of this later for it is a complex process. It involves balancing across the equation, but also making balances between the main classes of nutrients within the ration and then, further, within the nutrient groups, especially between the various amino acids.

In a sense it is like deriving sets of simultaneous equations in algebra and then solving them to get the appropriate proportions of the ingredients. In fact one important development of recent times is to accept this analogy and to allow the formulation to be completed by that very efficient solver of repetitive equations – the computer. Before this stage is reached, however, it is necessary to have a closer look at the questions set down above.

NUTRIENT REQUIREMENTS

The Agricultural Research Council of the United Kingdom (A.R.C.), and the National Research Council, of the United States (N.R.C.), publish booklets giving details of the nutrient requirements for most classes of poultry. These booklets have been accorded international acceptance and should be available in every formulation office.

Their recommendations are based on up-to-date reviews of the technical literature produced throughout the world on the subject. It is true that the bulk of such research has been conducted in countries within the temperate zones of the world and that, therefore, the statements of requirements are not always directly applicable to animals kept in other climates. However, more and more information is becoming available from countries with more extreme climates and, where they are desirable, modifications to the A.R.C. or N.R.C. standards can be made.

The most recently available recommendations are those published in 1975 by the Agricultural Research Council and reproduced here in Tables 16 and 17. Two points should be noted. First, the information is far from complete; many entries are italicized to show that they are tentative only and many blank spaces remain among the minerals and vitamins. Second, the requirements are set down in appropriate units per kg of diet and are linked thereby to an assumed daily intake of the ration (line 3 of Table 16).

What does this method of presentation serve to say? It does not of course, indicate the absolute requirement of the bird. It says rather that formulation should include a nutrient at the indicated rate if the known requirement is to be met.

It is important to realize that this approach is only valid if the estimate of the daily consumption of the ration is accurate. We have already noted the impact of departures from 'normal' intake on pages 45–47 and we must return to the point again, for in some environments, especially those with high ambient temperatures, food intake is bound to fall.

A similar argument is true in respect of the energy content of the ration and the A.R.C. tables assume an energetic value for the ration (e.g. 3·1 Mcal/kg in the case of young pullets). Of course, rations can be formulated readily with less energy per unit weight and fairly easily with more. So if some other ingredients need to be in balance with energy (e.g. for an optimum energy/protein

Table 16 Summary of the nutrient requirements of fowls, expressed as dietary concentrations. Figures in italics are tentative estimates only (A.R.C., London, 1975)

		0 to 4 weeks	*4 to 8 weeks*	*Growing pullets*	*Laying pullets*	*Breeding pullets*
Assumed ME content of diet	Mcal/kg	3·1	3·1			
	MJ/kg	13	13			
Assumed feed intake	g/day			75	110	110

Table 16 (*continued*)

		0 to 4 weeks	*4 to 8 weeks*	*Growing pullets*	*Laying pullets*	*Breeding pullets*
Crude protein	g/kg	*188*	*156*	*107*	*165**	*165**
Arginine	g/kg	10·3	7·6	6·0	5·1	5·1
Glycine + Serine	g/kg	14·0	10·2	8·0	?	?
Histidine	g/kg	4·8	3·6	2·8	1·7	1·7
Isoleucine	g/kg	8·5	6·4	4·9	5·5	5·5
Leucine	g/kg	14·7	10·7	8·4	6·8	6·8
Lysine	g/kg	11·0	8·0	6·3	7·5	7·5
Methionine	g/kg	4·8	3·6	2·8	3·5	3·5
Methionine + Cystine	g/kg	9·2	6·7	5·2	4·7	4·7
Phenylalanine	g/kg	8·5	6·4	4·9	3·9	3·9
Phenylalanine + Tyrosine	g/kg	15·8	11·6	9·1	7·0	7·0
Threonine	g/kg	7·4	5·3	4·2	3·6	3·6
Tryptophan	g/kg	2·1	1·5	1·2	1·7	1·7
Valine	g/kg	9·8	7·1	5·6	5·5	5·5
Calcium	g/kg	12	7·5	4	35	35
Non-phytin phosphorus	g/kg	4·7			3·5	3·5
Total phosphorus	g/kg		6·0	3·0		
Magnesium	g/kg	0·4	0·4	0·4	0·4	0·4
Sodium	g/kg	1·5	1·2	0·7	1·0	1·0
Chloride	g/kg	1·4	*1·1*	*0·6*	*0·9*	0·9
Potassium	g/kg	2·5	2·5	2·5	2·5	2·5
Copper	mg/kg	4				
Iodine	mg/kg	0·4				0·2
Iron	mg/kg	75				
Manganese	mg/kg	50	*50*	*50*	30	50
Zinc	mg/kg	40	*50*	*50*	*50*	60
Vitamin A	i.u./kg	1320	1320		2700	2700
Vitamin D_3	i.u./kg	400	400		600	600
Vitamin E	i.u./kg	*15*			?	?
Vitamin K_1	mg/kg	0·5			*1·0*	*1·0*
Thiamin	mg/kg	1·0			1·25	?
Riboflavin	mg/kg	4·0			2·5	4·0
Nicotinic acid	mg/kg	28			8	10
Pantothenic acid	mg/kg	10			1·5	6·5
Pyridoxine	mg/kg	3·5			2·0	4·0
Biotin	mg/kg	0·15				
Folic acid	mg/kg	1·5			0·3	0·5
Vitamin B_{12}	mg/kg	0·02†			?	0·002
Choline	mg/kg	1300			600	1100
Linoleic acid	g/kg	10	10	*10*	12‡	2§

* These protein and amino acid requirements are for an egg output of 50 g/hen day.
† Assumes negligible reserves at hatching.
‡ Requirement for maximum egg weight for pullets reared on conventional diets.
§ Requirement for maximum hatchability.

Table 17 Summary of the nutrient requirements of turkeys, ducks, geese, pheasants and Japanese quail, expressed as dietary concentrations. Figures in italics are tentative estimates only (A.R.C., London, 1975)

		Turkey poults	*Turkey growers*	*Breeding turkeys*	*Ducklings*	*Goslings*	*Pheasant chicks*	*Japanese quail*
Assumed ME content of diet	Mcal/kg	3·0	2·95		3·1	2·65		
	MJ/kg	12·6	12·3		12·8	11·1		
Crude protein	g/kg	280	220	*120*	230			
Arginine	g/kg	13·0	10·4		8·4	8·5		
Glycine + Serine	g/kg	9·0	7·2		11·3	11·4		
Histidine	g/kg	5·0	4·0		3·9	4·0		
Isoleucine	g/kg	9·0	7·2		6·9	7·0		
Leucine	g/kg	14·0	11·2		11·8	12·0		
Lysine	g/kg	13·0	10·4		8·9	9·0		
Methionine	g/kg	5·0	3·8		3·9	4·0		
Methionine + Cystine	g/kg	8·0	6·4		7·4	7·5		
Phenylalanine	g/kg	8·0	6·4		6·9	7·0		
Phenylalanine + Tyrosine	g/kg	14·0	11·2		12·8	13·0		
Threonine	g/kg	9·0	7·2		5·9	6·0		
Tryptophan	g/kg	2·2	1·8		1·7	1·7		
Valine	g/kg	10·0	8·0		7·9	8·0		
Calcium	g/kg	8	5	25	5·6	4	10	5
Non-phytin phosphorus	g/kg							
Total phosphorus	g/kg	6·5	5	5		4·6	8	4
Magnesium	g/kg	0·45	0·45	0·45	0·5			

Table 17 (*continued*)

		Turkey poults	*Turkey growers*	*Breeding turkeys*	*Ducklings*	*Goslings*	*Pheasant chicks*	*Japanese quail*
Sodium	g/kg	1·75	1·75	1·75			0·85	
Chloride	g/kg						1·1	
Potassium	g/kg	4·4	4·4	4·4				
Copper	mg/kg							
Iodine	mg/kg							
Iron	mg/kg	75						
Manganese	mg/kg	50		100			75	
Zinc	mg/kg	70					60	
Vitamin A	i.u./kg	1760		2700	?	?	?	
Vitamin D_3	i.u./kg	900		2000	300	300	1100	
Vitamin E	i.u./kg	*15*		*60*	?	?	?	
Vitamin K_1	mg/kg	1·5		?	?	?	0·06	
Thiamin	mg/kg	2·0		?	?	?	?	
Riboflavin	mg/kg	3·5		3·5	4·0	?	?	
Nicotinic acid	mg/kg	*35*		?	25	65	65	
Pantothenic acid	mg/kg	10·5		16	11	?	?	
Pyridoxine	mg/kg	4·0		?	2·5	?	?	
Biotin	mg/kg	0·25		0·2	?	?	?	
Folic acid	mg/kg	1·5		0·7	?	?	?	
Vitamin B_{12}	mg/kg	0·004		?	?	?	?	
Choline	mg/kg	1500		1350	?	?	1400	
Linoleic acid	g/kg			12				

ration) then it could be useful to specify ingredients not per unit weight of food but per unit of energy. However, the primary effect of altering the energy content of the ration is to modify daily intake and, providing an appropriate assumption can be made about intake (as in the A.R.C. tables), then requirements may be stated, as here, per unit weight of food.

One other aspect of accuracy is worth mentioning. Under commercial conditions it is necessary to consider a flock of birds as a unit and this means formulating to allow for varying levels of performance by the individuals within the flock and, of course, varying levels also of daily consumption of food. Thus the ration formulation should be based on a level of intake that will give the optimum performance for the most economical proportion of the flock. It should not be so high as to waste food on unproductive birds nor so low as to limit the production potential of any significant proportion of the flock.

Again, the nutrient requirements are based on research carried out at experimental stations where conditions are usually ideal. In the field it may be necessary to include margins of safety, particularly of the fat-soluble vitamins, to allow for loss of nutrients during processing and storage, variations in performance and for the various stresses to which poultry may be subjected.

Under normal conditions the margins of safety may be plus 50 per cent for the fat-soluble vitamins and plus 20 per cent for the water-soluble vitamins.

In the case of foods that have to be transported long distances in hot climates, higher margins still of safety may be required to guard against almost total loss of vitamin value at normal levels of inclusion.

VARIATIONS IN FOOD INTAKE

In a sense it is rather strange that modern statements of food requirements should be made in relation to an assumed constant daily food intake for, as we shall see, it has long been the practice to feed productive birds to choice (*ad libitum*), thus allowing them to regulate their total intake (and thus their intake of nutrients) themselves. Some types of laying chickens seem to be very good at this and eat in a very economical way. Others, especially strains of heavier bodyweight, show less control and, comparatively speaking, overeat to the point where production may suffer and profitability certainly does.

Because the cost of a bird's daily food intake is by far the highest single cost she bears, this issue has assumed overwhelming importance in determining profitability and great interest is now shown in methods of apportioning a daily allowance of food to a bird within quite sharp limits of accuracy (see Chapter Nine, page 208).

The daily food intake of light and medium hybrid layers varies considerably as Table 18 clearly shows:

Table 18 Variations in daily intake (g/bird/day) by brown and white pullets offered food *ad libitum*

		Weeks 25–28	*Weeks 21–48*	*Weeks 45–48*
Brown egg strains				
	1	116	122	132
	2	111	115	125
	3	116	116	124
	4	108	113	120
	5	116	119	127
	6	114	116	122
	7	120	122	128
	8	120	120	128
White egg strains				
	A	104	110	121
	B	109	111	122
	C	111	115	124
	D	92	97	102
	E	108	108	115
	F	109	111	121
	G	117	120	127
	H	106	107	123
	I	111	114	115

The figures show variations of 10 to 12 per cent at a time when birds are approaching peak rates of lay and when intake is likely to be of critical importance. It is also apparent that the intake of some strains increases regularly with time, that of others only quite late in lay and that of some scarcely at all. It is clear that a 'standard' ration is unlikely to be equally suitable for all strains all the time.

Many factors influence the voluntary food intake of laying birds. Among the more important are:

(i) Physiological mechanisms of the bird both as an individual and as a member of a strain;
(ii) The physical condition of the food;
(iii) The liveweight of the bird;
(iv) The rate of egg production;
(v) The nutrient density of the food mixture;
(vi) The environment in which the birds are kept.

Both liveweight and rate of lay have a causative effect on food intake. The relationship between the liveweight of the laying bird and its daily food intake follows from the fact that the larger the bird the more food is required for maintenance. Table 19 provides an illustration.

Table 19 Illustration of the difference in food intake of birds of different liveweights laying at the same rate

	Bird A	*Bird B*
Liveweight	1,800 g	2,500 g
Food intake:		
Maintenance	70 g	87 g
Production at 60 per cent	40 g	38 g
Total daily intake	110 g	125 g

The above example indicates that liveweight is an important economic factor in egg production. The additional 15 g of food eaten daily by the larger bird represents an additional annual consumption of 5·48 kg (about 12 lb).

A particular efficiency of some strains of bird is to adjust intake to level of production when there appears to be a clearly linear relationship between increased intake and increased production.

The physical condition of the food, that is whether it is provided in the form of mash or pellets, and the volumetric bulk of the diet, may also have a direct influence on daily intake. When given free choice, chickens eat rather more pellets than mash. Bulk is also important and if the food is too bulky, due to the presence in the mixture of fibrous ingredients such as oats, the birds may be unable to consume sufficient volume to maintain production. It is for this reason that the levels of inclusion of foods like bran, oats and dried yeast in the diet are limited.

Table 20 Illustration of the effect of rate of lay on food intake at constant liveweight

Rate of lay	60%	70%	80%	90%
Liveweight	1,800 g	1,800 g	1,800 g	1,800 g
Number of eggs	219	255·5	292	328·5
Food intake:				
for maintenance	70 g	70 g	70 g	70 g
for production	40 g	46·6 g	53·3 g	60 g
Total daily intake	110 g	116·6 g	123·3 g	130 g

So far it might seem that we are making the implication that a bird of a given weight and productivity will eat a characteristic weight of food each day, this amount varying from the norm only in respect of some variations in the physical condition of the diet. If the diet were formulated to a constant nutrient density this implication could be extended to suggest that within narrow limits the bird is able to monitor its nutrient intake and to stop eating when its physiological requirements have been met. This is, of course, an attractive hypothesis but, unfortunately, it does not quite fit the facts.

The physiological mechanisms that control the voluntary food intake of laying birds and which cause one bird or one strain to overconsume and become fat while others do not has been the subject of considerable research. It is generally considered that the centre in the brain responsible for monitoring appetite is the hypothalamus, a part of the endocrine system, but how the signals are passed and transcribed is not yet so well understood that voluntary appetite regulation could be a basis for formulating rations. The mechanisms are not sensitive enough.

Workers at Reading University have shown that almost all strains tend to increase their energy intake as the nutrient density of the diet is increased (even though they may have reduced food intake). Some do so to a greater extent than others. With strains of White Leghorns whose daily calorie intake ranged from 320 to 340 kcal the energy intake was shown to increase by 2 to 3 per cent for each 10 per cent increase in nutrient density; larger strains, consuming 360 to 400 kcal daily, increased by 4 to 5 per cent for each 10 per cent increase in dietary energy.

That some laying strains are by no means perfect, however, in their ability to limit their food intake according to their nutrient requirements and may, in fact overconsume, is well appreciated as is illustrated by the established practice of restricting the daily food intake of broiler breeding stock.

Daily food intake measurement is, therefore, an extremely important piece of information to have. In practice it is unusual to find that poultry farmers have any good measurement of how their flocks are faring in this respect. At best they may have some record of the amount of food that has gone into a poultry house (either in bags or in bulk) and have the ability to average this out over the number of birds and the days which the supply has lasted. This is not likely to be good enough if there are variations in the productivity through the house or if some of the feeding equipment permits substantial wastage of food (see Chapter Nine, pages 215–216).

By contrast, if intake is measured with some precision then feeding can be made both effective and economical. Thus the nutrient density of a ration designed for laying birds with an intake of 110 g per bird daily will need to be increased when the daily intake falls to 100 g, unless the margins of safety are designed to cover this eventuality. Alternatively the same rations fed to birds whose daily intake was 130 g would be overgenerous in nutrients and wastefully expensive.

NUTRIENT DENSITY

The term nutrient density or nutrient concentration is used to refer to the levels of the various important nutrients in a balanced food mixture. Initially it was commonly related to the energy content of the ration, but only because energy is the ingredient that can most readily be regulated in the diet. If the energy content is to be varied, for whatever reason, then it is necessary usually to vary the other ingredients proportionally.

Why then is nutrient density an important consideration? There are several reasons. The first is the important, if inexact, relationship (noted in the previous section) between density and intake. In so far as energy intake may tend towards a constant value (a tendency only, because natural regulation is not sensitive enough to ensure this) then total food intake falls as density rises, and increases as density is reduced. If, therefore, a given daily food intake is made the basis of ration formulation then there will also be an ideal density for that mixture. This assumption is clearly stated in the A.R.C. Summary Tables.

The second reason is also physiological and relates to the productive ability of the bird within the environmental context in which it is placed. A high level of production demands a given level of nutrient inputs. If the bird's capacity for receiving those inputs is limited, perhaps by body size, perhaps by a high ambient temperature, then increasing the nutrient density of the ration is the most obvious way by which the bird's productivity can be sustained. Indeed it was the advent of the small-bodied, egg-laying hybrid into modern commercial production that first focused attention on this particular aspect of poultry nutrition.

The breeders pointed out then that optimum performance depends upon the actual nutrient intake and that diet formulations must reflect the anticipated food intake of the new strains. The nutrient density must increase as intake is reduced. Accordingly the breeders publish guides to the food requirements of, and the methods of feeding for, their strains and these should be followed carefully when formulating rations.

The third reason is one of simple economics. The energy content of a ration varies in cost according to the availability and price of the major ingredients used to provide it. Thus for example maize, which is rich in its energy content, is readily available and comparatively inexpensive to use in formulations in the United States of America. A comparable ration made elsewhere, where maize is not grown, might not be economical, whereas a ration based on an indigenous cereal would be much less expensive though a good deal less rich in energy. In other words the cost of the ration per unit of energy has to be worked out. If it is much more economical to base a ration on ingredients which reduce its nutrient density then new assumptions concerning intake must be made. In general it is safe to reduce the density of a ration so long as the bird can eat more each day to obtain the calculated intake of nutrients. In this respect birds are capable of considerable adjustment. Thus experience shows that laying birds can give excellent production from a wide range of nutrient densities. The lower level of energy to meet the requirements of hybrid layers appears to be in the region of 2,310 kcal/kg while the higher the nutrient density the better will be the food conversion per egg produced (see Chapter Eight, pages 162–163). Usually the decision as to which nutrient density to use will depend, therefore, upon the availability of ingredients and the estimated cost of food per egg produced.

Of course, if diets of lower nutrient density are used and the bird eats more to keep up its nutrient intake, then more food has to be ordered, stored and put into the food troughs so higher total handling costs are incurred by feeding such rations. The general rule is that high-density ingredients and rations are more expensive to purchase but cheaper and more efficient to handle; low-density foods are cheaper initially but are often expensive to utilize and may actually limit the level of production that the birds can show.

FOOD CONVERSION EFFICIENCY

The question of cost utilization of food is an important one and is reflected in the frequently used measurement – food conversion efficiency, which calculates the amount of food used per unit product gained.

The producer's aim is to achieve the maximum rate of growth or the highest egg production at the lowest cost of food per unit of meat or eggs. Often the food that gives the best conversion ratio will also be the most economical to use, but not necessarily so. For this reason the food conversion ratio is interesting but not vital. It is the final unit cost of poultry meat or eggs that is of the greatest importance.

The food conversion ratio will deteriorate (i.e. more food per unit of produce) as the birds increase in liveweight and as egg production falls. One of the reasons for this is that the heavier the bird becomes or the less eggs it lays, the greater will be the proportion of food required for maintenance, and the smaller the proportion that is left for production purposes.

The more rapidly the bird grows or the more eggs it lays the better will be the food conversion ratio. Thus, strains of birds selected for rapid growth or for high egg production are more efficient in terms of food conversion than non-selected strains. Males of all species of poultry grow more rapidly than females and also require less food per unit of liveweight gain. For this reason they are preferred by the grower but not necessarily by the purchaser of the finished carcase.

So many factors influence the food conversion efficiency that it is difficult to compare one set of results with another on the basis of this factor alone. When ratios are used they should always be related to the liveweight of the birds and to their age and, in the case of egg production, to the rate of lay.

The impact of a knowledge of food conversion ratios is that it prevents the choice of ration being made on the basis of its cost per unit weight alone. It is quite possible to compile a ration that is of low cost per kilo yet most expensive when considered in relation to the amount and, therefore, cost of food per kilo of first-quality meat or eggs that it yields.

THE PROCESS OF RATION FORMULATION

The task of formulating rations has been likened to the process of solving a series of simultaneous equations. Once a statement of the bird (or flock) requirement has been set down, it is necessary to calculate the amounts of the nutrient sources available that can be blended together to give a serviceable and economic diet.

The process can easily be visualized if we make the (incorrect) assumption that all the energy required might come from one cereal alone, all the protein from a single protein source and so on. It would then be necessary to calculate only the amount of the cereal, the protein food and those supplying vitamins and minerals and then to measure these quantities out into the mixer.

In practice we know, of course, that a cereal also supplies protein, that a single protein source rarely offers all the necessary amino acids and that nearly all the major ingredients may contain some portion of the required vitamins and minerals. Consequently it is necessary to embark upon a process of adjustment and refinement of the formulation so that allowance is always made for the total contribution of each food source; amounts of each source are then increased or diminished until the requisite balances are achieved. This is a time-consuming process when it has to be done by trial and error, working out the content of so many substances in each ingredient contained in the ration.

The problem is increased many times as we remember that it is not only the nutritional balance in the formulation that must be achieved, but also that requirements covering the palatability, handling quantities and cost of the ration must also be satisfied.

Computer formulations

It is at this point that the computer comes into its own. It can resolve these long sequences of calculations in a negligible amount of time and can give a series of answers to each problem depending on limitations that may be set by the compounder in respect of any of the ingredients. That is to say that the computer will produce an answer that will be the formulation giving the lowest (or least) cost of ingredients. A few moments later it can offer another solution which takes into account the limitation that, say, fish meal must not comprise more than 5 per cent of the ration. A series of these limitations can be given to the computer which then supplies the best and least-cost answer after paying due account to each of them.

The computer is a most efficient calculating machine that must be fed, that is to say programmed, with the basic information in order to function. The efficiency of the computer is limited by the degree of accuracy of the information that is programmed and this can only be provided by a highly trained nutritionist. The computer is, therefore, a valuable support rather than a replacement for the nutritionist, who can now complete complex calculations in seconds that would have been impracticable if they had to be attempted without such a mechanical aid.

For devising formulations, computers make use of a principle known as linear programming. Linear programming is a mathematical formula, first developed in the early 1940s, that enables the optimum allocation of resources for a particular project to be achieved.

Programming the computer involves 'feeding' the following information into the machine:

(i) The nutrient content of the ingredients to be used;
(ii) The minimum and/or maximum levels to be included of certain ingredients;
(iii) The nutritional levels, i.e. the specifications required, in the formulations to be prepared;
(iv) The cost of the ingredients available.

The computer then calculates the 'least-cost', i.e. the most economical, formulation to meet the restrictions that have been demanded by the circumstances in which the compounder happens to be working at any one time. By the same token, the answer produced by the computer is relevant to the information, concerning those circumstances, which has been given to it. It will be properly used to find another answer as soon as the circumstances change.

It may not be necessary to program the computer to give information on all the nutrients that are required to make up the formula, but only for those that involve time-consuming mathematical problems. For example, many of the vitamins and trace minerals will usually be added to the mixture as a supplement containing known amounts of these nutrients and thus making the use of the computer unnecessary in respect of them.

The computer may be programmed to provide the formulation either as percentages or in unit weights of ingredients per kilo or per tonne.

Many companies manufacturing animal foods offer computer ration-formulation services, thus bringing the benefits of rapid and accurate least-cost formulation to the smallest organization. In addition a formulation service may be available from a university or government department.

Formulating rations without a computer

Calculating food formulations without the aid of a computer is very time-consuming and is likely to lead to an empirical, rather than a 'least-cost' formula being prepared.

The task may be simplified by using a protein, vitamin and trace-mineral concentrate, thus reducing the task of formulation to the best combination of cereals and perhaps of one vegetable protein-rich ingredient.

Special sheets may be prepared to ease the task of formulation and these may be of a simple nature for use when formulating for cereal and concentrate mixtures, or more complex for use when the formula is to be prepared without the use of a concentrate.

The levels of the minerals, calcium, phosphorus and salt must be balanced against the level of these nutrients supplied by the other constituents of the food mixture, in particular fish meal, meat and bone meal and the protein concentrates where these are used. The additional requirements for calcium and phosphorus will usually be supplied as limestone flour and di-calcium phosphate.

The quality of the calcium should also be checked regularly for inclusions of magnesium and that of the phosphorus supplement for fluorine.

If a computer formulation service is not available, it may well be best for the farmer to concentrate on the use of a few standard formulations that have been specially designed to a high nutrient standard. In this way the farmer is free to concentrate his experience on producing foods from a limited number of ingredients and in purchasing them in the most economical manner.

Ration formulation for egg production

SUMMARY OF FEEDING STANDARDS FOR INTENSIVE EGG PRODUCTION

Growing pullets

0–6 weeks of age	*Quoted range in formulation*
Type of food: chick starter	
Form of food: Mash or crumbs	
Crude protein content	19·5%–23·0%
Metabolizable energy density:	
kcal/kg	2,640–3,080
kcal/lb	1,200–1,400
6–14 weeks of age	
Food: Growers mash	
Crude protein content	15·5–18·5%
Metabolizable energy density:	
kcal/kg	2,640–3,080
kcal/lb	1,200–1,400
14–20 weeks of age	
Food: Pullet developer mash	
Crude protein content	12·5%–14·5%
Metabolizable energy density:	
kcal/kg	2,640–3,080
kcal/lb	1,200–1,400

Layers

20–42 weeks of age	
Food: Layers No. 1 mash (see subsequent notes)	
Crude protein content	14·5%–18·0% (moderate climates) 16·5%–20·0% (hot climates)
Metabolizable energy density:	
kcal/kg	2,640–3,190
kcal/lb	1,200–1,450

20–42 weeks of age

The quoted ranges should give the following nutrient intake figures:

Protein/g/hen/day	18 g
Energy/kcal/hen/day	305 (moderate climates) 275 hot climates

42 weeks onwards

Food: Layers No. 2 mash (see notes)	
Crude protein content	13·0%–15·5% (moderate climates) 14·0%–17·0% (hot climates)
Metabolizable energy density:	
kcal/kg	2,640–3,190
kcal/lb	1,200–1,450
Nutrient intake average:	
Protein/g/hen/day	16 g
Energy/kcal/hen/day	310 moderate climates 280 hot climates

SUMMARY OF FEEDING STANDARDS FOR BIRDS HOUSED SEMI-INTENSIVELY AND ON FREE RANGE

Growing pullets

0–8 weeks

Food: Chick starter mash, available at all times	
Crude protein content	19·5%–23·0%
Metabolizable energy density:	
kcal/kg	2,640–3,080
kcal/lb	1,200–1,400

8–20 weeks

Food: Growers mash available at all times; to encourage consumption a wet mash may be fed once or twice daily. Whole grain may be introduced at 6 weeks of age. When the birds are used to grain, 14 g (½ oz) of whole grain per bird may be given first thing in the morning and 40 g (1½ oz) per bird in the evening about one hour before dusk. Whole wheat or kibbled maize may be used. Insoluble grit should be provided once grain feeding is started.

Crude protein content	16·0%–17·0%
Metabolizable energy density:	
kcal/kg	2,640–3,190
kcal/lb	1,200–1,450

Layers

20 weeks onwards

Food: Grain balancer mash available at all times or with a wet mash once or twice daily, together with an early morning feed of grain (14 g or ½ oz per bird) and 40–45 g (1½–2 oz) per bird an hour before dusk. Whole wheat or kibbled maize may be used.

Crude protein content	17·0%–18·0%
Metabolizable energy density:	
kcal/kg	2,640–3,190
kcal/lb	1,200–1,450

Table 21 Summary of amino-acid requirements for layers to maintain production of 50 g of egg material/hen/day in (i) absolute intake in g per day and (ii) as a percentage of the diet at different levels of total daily intake of food

Amino acid	*Daily intake to produce 50g/egg hen/day*	*Assumed total daily food intake*				
		90g	*100g*	*110g*	*120g*	*130g*
	g	%	%	%	%	%
Arginine	0·56	0·62	0·56	0·51	0·47	0·43
Histidine	0·19	0·21	0·19	0·17	0·16	0·15
Isoleucine	0·61	0·68	0·61	0·55	0·51	0·47
Leucine	0·75	0·83	0·75	0·68	0·63	0·58
Lysine	0·83	0·92	0·83	0·75	0·69	0·64
Methionine	0·39	0·43	0·39	0·35	0·33	0·30
Methionine + Cystine	0·52	0·58	0·52	0·47	0·43	0·40
Phenylalanine	0·43	0·48	0·43	0·39	0·36	0·33
or Phenylalanine + Tyrosine	0·77	0·86	0·77	0·70	0·64	0·59
Threonine	0·40	0·44	0·40	0·36	0·33	0·31
Tryptophan	0·19	0·21	0·19	0·17	0·16	0·15
Valine	0·61	0·68	0·61	0·55	0·51	0·47

Table 22 Summary of amino-acid requirements of layers at different levels of production with eggs of 56·7 g (2 oz) and a daily food intake of 110 g/bird

Amino Acid	*100%*	*90%*	*80%*	*70%*	*60%*
Arginine	0·62	0·56	0·50	0·43	0·37
Histidine	0·21	0·19	0·17	0·15	0·13
Isoleucine	0·68	0·61	0·54	0·48	0·41
Leucine	0·83	0·75	0·66	0·58	0·50
Lysine	0·92	0·83	0·74	0·65	0·55
Methionine	0·43	0·39	0·34	0·30	0·26
Methionine + Cystine	0·58	0·52	0·46	0·41	0·35
Phenylalanine	0·48	0·43	0·38	0·34	0·29
Phenylalanine + Tyrosine	0·86	0·77	0·68	0·60	0·51
Threonine	0·44	0·40	0·35	0·31	0·26
Tryptophan	0·21	0·19	0·17	0·15	0·13
Valine	0·68	0·61	0·54	0·48	0·41

N.B. It will be noted that Table 22 relates to an egg weighing 56·7 g (2 oz), whilst Table 21 is calculated with reference to an output of 50 g egg material per day. An egg of 56·7 g is of normal commercial weight. Output is rarely 100 per cent whole eggs but, at 90 per cent of normal eggs, output is equivalent to a notional 50 g egg every day, which is, therefore, another more generalized way of describing average output. Estimates of rates of production are, however, usually made straight from numbers of eggs as in Table 22.

FORMULATION NOTES FOR LAYING BIRDS

Assumptions concerning productivity

Table 23 Egg production results (*Poultry Testing 1975*)

	WHITE EGG STRAINS		BROWN EGG STRAINS	
(i) Production				
140–532 days.	*Range*	*Average*	*Range*	*Average*
Eggs/Hen housed	211·3–292·7	265·1	245·6–286·3	266·4
Hen day production	59·8–76·4%	71·2%	61·1–74·9%	70·0%
(ii) Food conversion efficiency				
Kg food/kg eggs	2·50–2·90	2·67	2·67–2·86	2·77

Table 23 (*continued*)

	WHITE EGG STRAINS		BROWN EGG STRAINS	
(iii) Mortality	*Range*	*Average*	*Range*	*Average*
0–7 days	0–4·1%	1·1%	0–1·4%	0·3%
8–126 days	0–5·3%	1·2%	0–2·1%	0·7%
140–532 days	3·5–20·1%	9·9%	3·5–11·8%	6·6%

(iv) Housing
Rearing: deep litter, controlled environment
Adults: laying cages, controlled environment

Assumptions concerning management

(i) Stocking densities	*Birds per* m_2		*Sq ft per bird*	
Young stock:	*Floor*	*Cages*	*Floor*	*Cages*
0–4 weeks	21·5	63	0·5	0·17
4–8 weeks	10·8	31	1·0	0·34
8–20 weeks	8·1	21	1·5	0·50
Adults	4·5	14–21	2·5	0·5–0·75

(ii) Food hopper space	*Chickens on the floor*
First five days	One clean tray per 100 chicks
Growers	25–30 round feeders per 1,000 birds
Adults	40–50 round feeders per 1,000 birds
	Layers in cages
	10 cm (4 in) trough space per bird
(iii) Water supply	*Chickens on the floor*
	10 round troughs (37 cm) per 1,000 pullets
	Chickens in cages
	400 nipples per 1,000 pullets
(iv) Ventilation	11 m_3 minimum and 90 m_3 maximum of air per hour per kilo of food consumed daily

Assumptions concerning food intake

Attention has already been drawn to the importance of the absolute level of daily food intake. Table 24 indicates accepted levels in commercial production.

Table 24 Typical food consumption for commercial layers (*Poultry Testing 1975*)

		WHITE EGG STRAINS		BROWN EGG STRAINS	
(i) Total consumption		*Range*	*Average*	*Range*	*Average*
0–140 days	kg	7·57–9·02	8·53	8·43–9·39	8·98
141–532 days	kg	34·8–48·3	42·6	42·9–47·6	45·4
(ii) Daily intake					
141–532 days	g	98–126	115	115–124	119
(iii) Liveweight					
126 days	g	1,088–1,451	1,243	1,270–1,670	1,530
532 days	g	1,516–2,055	1,766	1,913–2,390	2,197
(iv) Housing					

Rearing: deep litter, controlled environment
Adults: laying cages, controlled environment

The aspect of the commercial production environment which will have the largest effect on intake is that of house temperature. As temperature rises the food intake declines at an ever increasing rate. The rate of decline might be for example, as the temperature changes from 20° to 30° C, 1·5 per cent per °C rise, whereas from 30° to 38° C the fall may be 4 to 5 per cent per °C.

At house temperatures above 32° C considerable difficulty may be experienced in maintaining intake at a level sufficient for maintenance and production. As ambient temperature approaches normal body temperature (41·7° C) the birds may at first utilize body reserves as a substitute for intake. Closer still to body temperature, food, intake and production can be expected to collapse completely.

In respect of dietary calorie intake, this tends to fall by about 4 kcal/bird/day per °C rise in temperature, but the rate is slower at low temperatures and much faster at high temperatures. At 20° C high-producing birds may require about 320 kcal of metabolizable energy daily, whereas at 30° C the figure may fall to 280 kcal.

However, the requirements for nutrients other than energy remain fairly constant. In very hot weather, therefore, it is necessary to increase the level of these nutrients as a percentage of the anticipated daily food intake, in order to maintain a constant level of consumption. An example of how this may be done is shown in Table 25.

It is seldom practical to make frequent changes in the formulation to match it to the house temperature. One attempt at a practical solution is to prepare two standards of formulations, one for moderate temperatures and one for periods of high temperature.

Table 25 Illustration of the approximate manner in which the nutrient density of a layer's diet should increase when the intake falls as the ambient temperature rises

Temperature	20 °C	24 °C	27 °C	30 °C
Food intake daily	110 g	105 g	100 g	95 g
Metabolizable energy: kcal/day/bird	320	300	287	270
Egg production: 50g/eggs/hen/day				
Metabolizable energy: kcal/kg	2,900	2,860	2,870	2,840
Crude protein, 18 g/bird/day %	16·5	17·14	18·0	18·95
Calcium %	3·5	3·67	3·85	4·05
Non-phytin phosphorus %	0·35	0·37	0·39	0·41
Sodium %	0·10	0·105	0·11	0·12
Arginine %	0·51	0·53	0·56	0·59
Histidine %	0·17	0·18	0·19	0·20
Isoleucine %	0·55	0·58	0·61	0·64
Leucine %	0·68	0·71	0·75	0·79
Lysine %	0·75	0·79	0·83	0·87
Methionine %	0·35	0·37	0·39	0·41
or				
Methionine + Cystine %	0·47	0·50	0·52	0·55
Phenylalanine %	0·39	0·41	0·43	0·45
or				
Phenylalanine + Tyrosine %	0·70	0·73	0·77	0·81
Threonine %	0·36	0·38	0·40	0·42
Tryptophan %	0·17	0·18	0·19	0·20
Valine %	0·55	0·58	0·61	0·64

The margins of safety for vitamins and trace minerals will usually be sufficient to allow for changes in daily food intake but nevertheless the levels should carefully be checked.

The optimum temperature for laying birds to allow the most efficient food conversion and production would appear to be between 25° and 30° C (75° to 85° F) whilst a minimum of 18° to 20° C and a maximum of 32° C cover the range for good average production.

It may be cheaper in temperate climates to spend money on heating or insulating the houses, rather than to allow the chickens to consume extra calories to maintain body heat.

In areas of high temperatures and particularly when high temperature is accompanied by high humidity, the economic problem is to decide whether to spend money on air-conditioning equipment or whether it is better to produce the eggs in a more equable climate and transport them to the site where they are needed.

Energy intake

Energy intake is not likely to be critical to production for intensively managed birds in temperate climates. Only if the main energy source contemplated for the ration (e.g. oats) is itself of low energy density may the bird be unable to eat enough energy for high production. Even in this example birds can still ingest enough energy provided that the ration is pelleted.

Provided then that the energy costs of the various energy contributors (the major cereals) are roughly equivalent it is possible to recommend that the ration is planned to give an intake of 300 to 320 kcal of energy/bird/day. This is why the A.R.C. recommendations are based on an assumed intake of 310 kcal/day. If energy costs are not comparable and one cereal is disproportionately cheaper to include in the ration, departures of 10 per cent from this figure could be contemplated without serious hardship to the bird. A comparable basal intake level for a heavier-bodied brown-egg strain might be 360 kcal/day.

Some strains of bird are known to eat substantially more than they should and may be deliberately restricted. The programme recommended by the breeder for such birds should be followed; this may involve specially designed formulations for restricted feeding.

Protein content

The figures given on pages 156–157 for crude protein content and intake must be related to the discussion in Chapter Three. The production response of laying birds to crude protein input is curvilinear and an intake of 16 to 18 g/day corresponds broadly to the point on the curve where increased production may be too little to warrant the increased expenditure on raising the protein content of the ration (and hence protein consumption) still further. Modern strains of laying bird continue to show sizeable responses to intakes of 20 to 22 g/day and at times, in the rise to peak production and the early weeks after peak, this level could perhaps be justified for a substantial proportion of the flock. Again it is the cost of the extra protein intake that is likely to be the critical factor controlling the formulation decision. Currently work is in train to discover the levels of intake of individual amino acids which must control response at around this level of intake. Recommendations, therefore, may change as this knowledge and the availability of single amino-acid sources improves.

In the same sense it is fair to say that an intake of 16 g of crude protein per day is the minimum for high-level production given normal protein sources for inclusion in the ration. This is what promotes the differences set out on pages 156–157 under the heading Layers No. 1 and Layers No. 2 mashes. In early lay there is almost certainly good evidence for encouraging higher levels of intake than 16 g/day. Later in lay the bird's performance does less to justify this recommendation. She becomes less efficient in the task of protein utilization and the response course of older birds offers less promise for higher inputs as a result. It is possible, therefore, to specify a daily protein intake for the second half of lay of 16 g/day. This reduction in standard is welcomed commercially because it makes for a cheaper ration. The important point to make is that this cheapening of the ration as lay proceeds can be as dangerous as it is attractive. The bird's requirement for protein is not reduced in any systematic way – if anything it increases because of failures of utilization. Only when it is clear that the slope of the flock's response curve has changed should these savings be considered. If they are taken too early an undesired limitation on production may well be set.

In summary, then, we may say that the daily intake of protein of high-producing birds may range from 16 to 20 g and the requirements remain fairly steady throughout the year. The reason for this is that as production declines the weight of individual eggs tends to increase, thus the total weight of egg output remains fairly constant.

CALCIUM

It is usual to include all the supplementary calcium in the food mixture. In very hot weather it may be necessary to supply additional calcium to laying birds, either as oyster shell or limestone grit, spread on the food in the hoppers. Granular limestone should be used if very dusty food results from the use of limestone flour. The quality of the limestone should be checked for magnesium content as a matter of routine.

Table 26 Approximate calcium requirements in layers' foods according to food conversion

G of food/egg	*Lb/Food/doz. eggs*	*% Ca./Diet*
120	3·16	3·75
130	3·43	3·46
140	3·69	3·21
150	3·96	3·00
160	4·22	2·81
170	4·49	2·65
180	4·75	2·50
190	5·02	2·36
200	5·28	2·25

The table has been prepared on the assumption that a 57 g egg (2 oz) contains approximately 2 g of calcium, that this is obtained from the calcium in the food and from the reserves in the bones and that, of the calcium consumed, less than half is retained whilst the remainder is excreted. Thus 4·5 g of calcium must be consumed to lay one 57 g egg (equals 0·12 lb of calcium to lay one dozen 2 oz eggs).

Phosphorus

All the phosphorus from animal sources and not more than 40

per cent of the phosphorus from plant sources may be regarded as being available.

Likely poisonous substances

Some common ingredients have occasional (or regular) unpleasant contaminants. Thus:

Cottonseed meal Decorticated cottonseed meal should be checked for gossypol. Use four parts of iron sulphate to one part of gossypol, up to 400 ppm, to avoid the problem of discoloured yolks.

Groundnut meal Groundnut meal must be checked for aflatoxin and only samples used that are free from this toxin.

Coccidiostat

A coccidiostat should be included in all starter foods designed for chickens on litter, and may be required for pullets and adult birds on other systems, as advised by the veterinarian.

NUTRITION AND EGG QUALITY

Egg size

Egg size is usually considered in terms of egg weight and this is influenced by many factors. These include:

(i) Level of protein in the diet;
(ii) Mineral levels;
(iii) Vitamin levels;
(iv) Level of linoleic acid;
(v) Influence of strain.

Protein has an immediate effect on egg weight and within the range 14 to 20 per cent crude protein, rations that have been carefully balanced in respect of amino acids give heavier eggs the higher the protein level.

The choice of protein level in the layers' diet, therefore, may depend on an accurate evaluation of the extra cost for the additional protein compared with the extra income from the larger eggs obtained. Light hybrids laying at a high rate often produce large numbers of very small and unsaleable eggs in the early stages of lay. For these, anything that increases the number of saleable eggs may be economic especially if there is a market premium for larger eggs.

Approximately 1 per cent additional large eggs may be anticipated for every 1 per cent increase in the crude protein content of the diet up to a maximum of 20 per cent crude protein.

It has been shown that the requirement for dietary calcium varies with the level of vitamin D_3 in the diet and that a high calcium level in the diet, with a low level of vitamin D_3, may result in a significant decrease in egg weight.

The unsaturated fatty acid linoleic acid has been shown to be one of the factors leading to higher egg weight. Linoleic acid is found in vegetable oils and maize has one of the highest natural contents among the common ration ingredients. The effect of linoleic acid is on egg size and not on egg numbers.

Egg size is also influenced by the strain of the chicken. Breeders usually publish information concerning the average egg size that can be expected from their particular strains.

Shell quality

The quality of egg shells is intimately linked with the presence in the diet of adequate levels of vitamin D_3 and certain minerals including calcium, magnesium, phosphorus and possibly zinc.

There is a relationship between the level of vitamin D_3 and the minerals calcium and phosphorus and a deficiency or imbalance of one or other of these nutrients may result in a decrease in shell thickness and misshapen eggs followed by reduced egg production. The production of thin-shelled eggs is one of the first symptoms of vitamin D_3 deficiency.

Thin and brittle-shelled eggs have been associated with a deficiency of manganese. An excess of dietary magnesium above 0·2 per cent may result in a loss of shell thickness.

A frequent cause of poor-quality egg shells during periods of very hot weather is the failure to increase the nutrient density of layers' food in respect of the nutrients other than energy.

Egg-shell quality also falls towards the end of the laying period. This appears to be due to a failure in calcium metabolism and is not remedied by increasing the calcium content of the ration.

Other nutritional factors are important. The use of certain sulphonamide drugs may lead to the laying of thin-shelled eggs. The feeding of seed grains that have been treated with certain insecticides and fungicides has been shown to be responsible for the production of malformed eggs.

The use of rancid cod liver oil in the diet has been reported to be responsible for the appearance of eggs with rough shells, showing an absence of natural glossiness.

The production of eggs with poor shell quality may also be due to the presence of disease even at a low level, e.g. infectious bronchitis and Newcastle disease.

Internal egg quality

The nutritive content of the egg in respect of many of the important trace minerals and vitamins depends upon the level of these nutrients in the diet of the laying hen. For example, the iodine content of the egg can rapidly be increased by the addition of suitable iodine supplements to the diet.

In the same way the level of riboflavin in the diet is reflected in the level of riboflavin found in the egg. In sufficient quantities riboflavin imparts to the albumen a very slight yellowish-green tinge.

The incidence of blood spots in eggs does not seem to have any obvious nutritional cause. Although alfalfa, vitamin K and iron salts have been implicated from time to time there is no experimental evidence to suggest the connection.

Yolk colour

The colour of egg yolks, however, does depend upon the presence in the ration of carotenoid pigments known as xanthophylls. The pigments are found in fresh and good-quality dried greenfood and in addition pigments can be purchased and included in the ration as a food additive. The subject of pigments has been dealt with in Chapter Six.

When deep-yellow yolks can be sold at a premium one of the following ingredients should be included in the formula:

(i) 30 per cent of yellow maize;
(ii) 5 per cent of good-quality dried grass or alfalfa;
(iii) Up to 22 mg/kg (10 mg/lb) of xanthophyll.

By the same process undesirable abnormally coloured yolks may also result from the bird's consumption of certain highly pigmented plants or by the use of unsuitable foods.

Cottonseed meal may contain constituents (one of which is a toxin known as gossypol) which, when consumed by laying birds, may result in the production of olive-green to dark-brown yolks, and in a pinkish discoloration of the albumen in eggs that have been cold stored.

Olive-coloured yolks may also result from the heavy consumption by free-range hens of acorns (*Quercus robur*), shepherd's purse (*Capsella bursa-pastoris*), field pennycress (*Thalaspis arvense*), or from the consumption of silage.

The inclusion of pimento pepper in the diet of laying birds leads to the production of eggs with orange-red yolks. In addition a number of fat-soluble dyes, e.g. Sudan 2, can be readily transferred to the yolk through the hen's diet.

LAYERS' FORMULATION CHECK LIST AND SPECIMEN FORMULATION

Table 27 Information required from the farm or flock

1	Name of farm;	7	Level of production;
2	Flock number;	8	Daily food intake;
3	System of housing;	9	Grams of food per egg produced;
4	Strain of chickens;		
5	Number in flock;	10	Liveweight of birds;
6	Age;	11	House temperature;
		12	Feeding programme.

FORMULATION SEQUENCE

1 Decide the nutrient levels to meet the percentage egg production and daily food intake.

Table 27 (*continued*)

FORMULATION SEQUENCE

2 Decide the restrictions of levels of use of ingredients due to the price and other considerations.
3 Formulate alternative rations.
4 Compare the formulations on the basis of cost per tonne per 1,000 kcal.

SCHEMATIC COMPUTER FORMULATION OF A LAYERS' DIET

1 Specifications made by poultryman

(i) Intake per bird:	108–115 g/day
(ii) Minimum standards of nutrient composition:	
Energy level	2,772 kcal ME/kg
Crude protein	15·2%
Lysine	0·7%
Methionine	0·32%
Calcium	3·4%–3·5%
Phosphorus (available)	0·4%
(iii) Maximum standards:	
Salt	0·4%
(iv) Guidance targets:	
Linoleic acid	1·5%
Pigmenting xanthophylls	11 mg/kg

2 Feeding stuffs available and ruling prices

	£/*tonne*
Maize	68.00
Wheat	65.00
Barley	64.00
Middlings	62.00
Soya-bean meal	93.00
Herring meal	158.00
Meat and bone meal (50% CP)	97.50
Grass meal	66.00
Pigment premix (potency 3·45 mg/kg)	1.20 per kg
Di-calcium phosphate	105.00
Limestone	10.00
Methionine	0.95 per kg

For salt and mineral/vitamin supplement assume 75p/tonne of finished mixture.

Table 27 (*continued*)

SCHEMATIC COMPUTER FORMULATION OF A LAYERS' DIET

3 Formulation

Four formulations have been produced sequentially for comparison, according to the following sources of variation:

Formula 1:

Notes (i) No synthetic methionine offered. This leads to a large inclusion of herring meal.

(ii) Maximal inclusion rates set for

Wheat	at 40%
Middlings	at 15%

Formula 2:

Notes (i) Synthetic methionine made available. Leads to dramatic fall in use of herring meal, whose residual protein value is now made up by soya-bean meal.

(ii) From Formula 1 it is noted that the combination of grass meal and maize is a cheaper source of pigment than a special concentrate. So the grass meal inclusion is now fixed at 2·5 per cent.

(iii) Constraints retained for wheat and middlings and a *minimum* level for herring meal is set at 1·25 per cent.

Formula 3:

Notes (i) The price of wheat is assumed to have risen to £67.50 per tonne, at which level it becomes too expensive to include.

(ii) Maize and wheat are close to being interchangeable for nutrients and are readily balanced with middlings and, in this case because of the constraint on middlings, a little barley.

Formula 4:

Notes (i) No grass meal is available but wheat price has returned to £65.

(ii) Maize is a cheaper source of pigment than premix whilst it and middlings replace wheat and soya and whilst grass meal is available. Pigment premix becomes cheaper if grass meal is not available and a maximum level of middlings has been reached.

Table 27 (*continued*)

SCHEMATIC COMPUTER FORMULATION OF A LAYERS' DIET

Formula	*1*	*2*	*3*	*4*
Ingredient	%	%	%	%
Maize	27·7	30·7	58·5	43·3
Wheat	40·0	40·0	—	20·1
Barley	—	—	1·9	—
Middlings	12·2	4·5	15·0	15·0
Herring meal	7·1	1·2	1·2	1·2
Soya-bean meal	—	8·6	8·5	8·0
Meat and bone meal	2·5	4·6	4·4	4·4
Grass meal	2·5	2·5	2·5	—
Limestone	7·6	7·5	7·6	7·6
Di-calcium phosphate	—	—	—	—
Methionine	—	0·052	0·056	0·058
Pigment premix	—	—	—	0·04
Salt and vitamin premix	0·4	0·4	0·4	0·4

These rations yield the following nutrient characteristics:

Formula		*1*	*2*	*3*	*4*
ME (kcal/kg)		2,772	2,772	2,772	2,772
Crude protein	%	15·5	15·4	15·2	15·2
Oil	%	3·1	2·6	3·5	3·1
Fibre	%	3·2	3·5	4·1	3·6
Lysine	%	0·77	0·70	0·71	0·70
Methionine	%	0·32	0·32	0·32	0·32
Methionine + Cystine	%	0·57	0·59	0·57	0·58
Calcium	%	3·5	3·5	3·5	3·5
Available phosphorus	%	0·4	0·4	0·4	0·4

They have the following comparative costs:

	1	2	3	4
Ingredient cost £	68·42	66·89	67·31	67·21
(+ supplement)	0·75	0·75	0·75	0·75
	69·17	67·64	68·06	67·96

Note: These are least-cost solutions and run very closely with the ruling prices of the major cereal ingredients.

Ration formulation for broiler growing

SUMMARY FEEDING STANDARDS FOR BROILERS AND CAPONS

0–3/4 weeks	*Quoted range in formulation*
Type of food: Broiler starter	
Form of food: No. 1 crumbs	
Crude protein	20·5%–25%
Metabolizable energy:	
kcal/kg	2,750–3,300
kcal/lb	1,250–1,500
4–8/9 weeks	
Type of food: Broiler finisher	
Form of food: No. 2 crumbs or 3·2 mm pellets	
Crude protein	17·5%–21%
Metabolizable energy:	
kcal/kg	2,860–3,410
kcal/lb	1,300–1,550
*8/9–12/16 weeks**	
(Capons only)	
Type of food: Broiler finisher No. 2	
Form of food: 4·7 mm pellets	
Crude protein	15·5%–18·5%
Metabolizable energy:	
kcal/kg	2,600–3,080
kcal/lb	1,200–1,400

FORMULATION NOTES

Assumptions concerning productivity

Appetite in broilers is perhaps even more critical than in layers. In fact growth rate may be regarded as a function of food consumption since the liveweight of a broiler increases in proportion to the amount of food consumed over and above that needed for maintenance. For this reason a certain quantity of food will not lead to the same increase in bodyweight at any age for, as the bodyweight increases, so does the amount of food required for

* A synthetic hormone implantation may be given at 3/4 weeks prior to killing (see formulation notes).

maintenance. From a fixed intake, therefore, less is available for growth. To maintain growth, intake must increase but even if it does each unit of growth bears an increasing share of the food costs of maintenance. Thus a 2·0-kg broiler requires considerably more food per unit liveweight than a 1·5-kg broiler.

Table 28 Standards for broilers by sex and as hatched (*Hubbard 1974*)

AGE	AVERAGE WEIGHT	FOOD CONVERSION (cumulative)	APPROXIMATE KG FOOD REQUIRED PER 1,000 BROILERS		
Weeks	*Kg*		*Daily*	*Weekly*	*To date*
As Hatched					
1	0·12	0·91	15	108	108
2	0·31	1·16	36	250	358
3	0·51	1·30	43	303	660
4	0·76	1·43	61	429	1,089
5	1·04	1·58	80	559	1,648
6	1·34	1·78	105	733	2,382
7	1·63	1·94	112	786	3,168
8	1·93	2·07	117	823	3,991
9	2·22	2·18	122	855	4,845
Males					
1	0·13	0·88	16	116	116
2	0·34	1·12	38	265	381
3	0·56	1·26	46	323	704
4	0·84	1·39	66	463	1,166
5	1·14	1·53	83	583	1,749
6	1·45	1·73	109	762	2,511
7	1·78	1·88	119	832	3,343
8	2·11	2·01	128	897	4,240
9	2·45	2·12	136	953	5,193
Females					
1	0·10	0·95	14	99	99
2	0·28	1·21	34	235	335
3	0·46	1·35	40	283	617
4	0·68	1·48	57	396	1,013
5	0·94	1·64	76	535	1,548
6	1·22	1·84	101	705	1,153
7	1·49	2·01	106	740	2,993
8	1·75	2·14	107	749	3,742
9	1·99	2·25	108	756	4,498

When comparing the relative efficiency of different batches of broilers that have been marketed at the same time, it is essential to consider this important relationship existing between food consumption and liveweight. The performance efficiency ratio of the flock is an attempt to allow such a comparison to be made, and is obtained by multiplying the liveweight of the birds by 100 and dividing the result by the food conversion ratio. The food conversion ratio is calculated by dividing food consumption by liveweight.

Expected growth rates and food conversion figures for well-grown broilers are set out in Table 28.

Occasional reports of loss of appetite are encountered on broiler plants and immediate action is necessary to restore food intake. A check on the possibility of the presence of coccidiosis should be made and treatment given as necessary. One of the possible causes of loss of appetite is the presence of a non-specific infection. In such cases encouraging reports have been made regarding the use of high-level antibiotics, or following the use of a vitamin-trace mineral supplement.

Apparent lack of appetite may be due to a changeover being made rather suddenly from one size of feeder to another. The chickens become accustomed to eating from one type of hopper; when a change is made, the new food hoppers should be placed in the house for a few days before the old ones are moved to make sure the birds are really used to them.

Changes from one food mixture to another should be made gradually, for otherwise the chickens may decline to eat the new mixture for a few days. Food intake may easily fall if stale food accumulates in the hoppers. Fresh food is most important at all times.

It is to be noted in Table 28 that figures are given separately for males and females. Males grow more rapidly than females and their food conversion is also superior. As a result, when a batch of broilers of mixed sex is killed there is considerable variation in size both between and within sexes.

Rearing male and female broilers separately enables birds of more uniform size to be produced and may also allow some food savings to be achieved.

Assumptions concerning management standards

Stocking density The stocking density for broilers and capons may be calculated on the basis of a stocking rate of 29·3 kg liveweight p/m_2 (6 lb/sq ft) for broilers raised in fully controlled environment houses. In houses that are less well insulated and ventilated, additional floor space should be allowed.

	Liveweight at disposal (kg)	*Birds/m²*	*Sq ft/bird*
As hatched	1·9	15·0	0·70
Males	2·1	14·0	0·77
Females	1·7	17·0	0·63
Capons	2·7	10·0	1·00

Food hopper space 25 to 30 round feeders or pans per 1,000 broilers.

Water supply Ten round automatic troughs per 1,000 birds.

Ventilation 11 m³ minimum, 90 m³ maximum of air per hour per kg of food consumed daily.

Pre-starter

Many broiler producers use a so-called pre-starter broiler mixture of very high nutrient density during the first seven to ten days. Usually the food is used at the rate of 255 kg (5 cwt) per 1,000 broilers followed by the normal broiler starter.

Fats

The addition of animal fat or vegetable oils at the rate of up to 3 per cent of the food mixture will be of considerable help in reaching the high energy levels suggested; it will also help incidentally to reduce the problem of dusty food.

Biotin

In cases where large quantities of wheat are to be used, the recommended levels of biotin should be thoroughly checked as a precaution against outbreaks of 'fatty liver kidney syndrome'.

Pigmentation

For a market where deeply pigmented yellow skin colour is prized, maximum levels of yellow maize should be used in the ration and, possibly, a pigment included in the food mixture for the last four weeks prior to killing, to supply up to 22 mg/kg (10 mg/lb) of pigmenting xanthophylls (see Chapter Six, pages 103–104).

Growth promotant

Generally more economic results will be achieved if broiler mixtures contain the recommended level of a suitable growth promotant.

Coccidiostat

A coccidiostat should be included in all broiler rations designed for birds reared on litter and may be required for broilers reared on other systems.

Insoluble grit

The supply of insoluble grit is important if coarsely ground foods are fed. In the absence of grit some of the grain may pass completely through the birds undigested so that food conversion ratios deteriorate.

A little chick-size grit may be sprinkled on the food during the first and second weeks after which insoluble grit may be given every two weeks. Medium-sized grit may be used after four weeks and adult size after six weeks.

When high-efficiency broiler rations are used in the form of well-made crumbs or pellets, the use of insoluble grit should be unnecessary.

Consumption of grit per 5,000 birds is likely to be 25 kg of chick-size grit, 25 kg of medium-size, and 25 kg of coarsely ground grit up to ten weeks of age.

Wet litter

Wet litter is a problem that is sometimes encountered in broiler production and, although the problem is usually associated with husbandry and management, there are some nutritional aspects that must be considered.

Wet litter, if due to nutritional causes, is likely to follow from excessive consumption of water, itself the result of a ration which contains too much salt. A high salt content may be caused by the use of very high levels of foods rich in animal protein.

Another possible nutritional cause of wet litter is the inclusion of ingredients that have become rancid. The use of maize meal that has been ground and stored for a long period may cause trouble, and it is for this reason that the use of freshly ground maize in broiler diets is essential.

Yet again the inclusion of a high level of ingredients that tend to have a laxative effect, e.g. molasses, may be the cause.

Nutritional reasons aside, inadequate ventilation is probably the prime cause of wet litter. Moisture from the bird's respiration is not removed and the air is too humid. If this is coupled with unsatisfactory water receptacles and poor litter management, the litter will never be good; nor will the performance of the broilers using it.

Caponization

If birds are required to grow to weights above 2 kg, they are usually caponized. Details of the process are set out on pages 99–100.

Specimen formulations for broiler diets

Ingredient		*Broiler starter*	*Broiler grower*
Maize	%	25·0	12·5
Wheat	%	46·3	55·0
Barley	%	—	5·2
Herring meal	%	10·0	7·5
Soya-bean meal	%	12·5	13·7
Meat and bone meal	%	2·5	2·5
Dried distillers' solubles	%	1·2	—
Fat	%	1·0	2·2
Methionine	%	0·064	0·056
Limestone	%	1·0	0·9
Salt and vitamin supplement	%	0·5	0·5
Coccidiostat		+	+

These yield:

		Broiler starter	*Broiler grower*
ME	kcal/kg	3,080	3,080
Crude protein	%	21·50	20·40
Lysine	%	1·20	1·10
Methionine	%	0·50	0·45
Methionine + Cystine	%	0·84	0·79
Oil	%	4·10	4·80
Fibre	%	3·00	3·10
Calcium	%	1·00	0·90
Available phosphorus	%	0·46	0·43

Ration formulation for breeding chickens

SUMMARY FEEDING STANDARDS FOR REPLACEMENT PULLETS AND BREEDING STOCK

0–6 weeks	*Quoted range in formulation*
Food: Chick starter mash	
Crude protein content	19·5%–23%
Metabolizable energy:	
kcal/kg	2,640–3,080
kcal/lb	1,200–1,400

6–14 weeks	
Food: Growers' mash	
Crude protein content	15·5%–18·5%
Metabolizable energy:	
kcal/kg	2,640–3,080
kcal/lb	1,200–1,400

14–20 weeks	
Food: Pullet developer mash	
Crude protein content	12·5%–14·5%
Metabolizable energy:	
kcal/kg	2,640–3,080
kcal/lb	1,200–1,400

20 weeks onwards	
Food: Breeders' mash	
Crude protein content	14·5%–17·0% moderate climates
	16·0%–19·0% warm climates
Metabolizable energy:	
kcal/kg	2,640–3,190
kcal/lb	1,200–1,450

FORMULATION NOTES

Assumptions concerning performance of layer-type breeders

Table 29 Typical commercial layer–breeder results (*based on ADAS data*)

Production	WHITE EGG STOCK	BROWN EGG STOCK
Total hatching eggs per day-old breeder pullet	160–190	160–210
Total hatching eggs per breeder housed	166–197	166–218
Live chicks per breeder	132–157	128–168
Age in weeks		
First egg	20–24	21–25
10% production	24–28	25–29
80%–90% production	26–32	28–32
50%–55% production	62–70	61–69
Food consumption (including food for males)	*kg/bird*	*kg/bird*
Chick mash	1·27– 1·45	1·50– 1·68
Growers' mash	6·35– 7·26	7·26– 8·16
Breeders' mash	35·38–38·56	37·20–43·10
Total	43·00–47·27	45·96–52·94
Liveweight kg		
24 weeks	1·13–1·59	1·59–2·04
65 weeks	1·54–1·91	2·00–2·36
Mortality and culls		
0–24 weeks	4·0%–7·0%	4·0%–7·0%
24–65 weeks	3·0%–8·0%	3·0%–8·0%

Table 30 Typical commercial broiler–breeder results (*based on ADAS/NFU Broiler-Breeder Bulletin 1975*)

Total hatching eggs per day-old breeder pullet	120–150
Total hatching eggs per breeder housed	126–158
Live chicks per breeder	84–135

Table 30 (*continued*)

Age in weeks	
22–24	first egg
26	10%
30–32	75%–80%
60–61	35%–45%

Food consumption, 0–61 weeks

	lb/bird	*kg/bird*
Chick mash	3·0	1·36
Growers' mash	12·6	5·72
Breeding mash	91·8	41·64
Oats	15·6	7·08
Flint grit	0·5	0·23
Total	123·5	56·03

Liveweight		*lb*	*kg*
24 weeks	females	4·25	1·92
61 weeks	females	5·50	2·49

Mortality and culls	*females*	*males*
0–24 weeks	5%	39%
24–61 weeks	8%	10%

Assumptions concerning management standards

Stocking density

	Birds per m²	*Sq ft per bird*
Layers	4·0	2·7
Broiler breeders*	3·5	3·0

Feeding space 40 to 60 round feeders or pans per 1,000 birds.
Water supply Six round automatic troughs per 1,000 birds.
Ventilation 11 m³ minimum and 90 m³ maximum of air per hour per kilo of food consumed daily.

Nutrient levels

Breeding birds require higher levels of certain vitamins and

* With skilled practice this may be reduced towards 5 birds per m² (2·2 sq ft per bird).

minerals than laying birds (see Table 31) and the levels should be checked.

Table 31 Increase in requirement of certain nutrients by breeding birds compared with laying birds (*A.R.C. 1975*)

	mg per kg of diet	LAYERS	BREEDERS
Vitamins:			
	B_{12}	?	0·002
	Choline	600	1,100
	Folic acid	0·3	0·5
	Niacin	8	10
	Pantothenic acid	1·5	6·5
	Pyridoxine	2·0	4·0
	Riboflavin	2·5	4·0
Minerals:			
	Iodine	—	0·2
	Manganese	30	50
	Zinc	50	60

Amount of food to be made available

The amount of food to be supplied should be in accordance with the instructions of the breeder who is likely to recommend a restricted feeding programme, especially for broiler breeders.

Coccidiostat

A coccidiostat will be required in the starter food for birds reared on litter and may be required for birds reared on other systems and during the growing stage.

Alternative feeding

In the event of hatching eggs not being required from the breeding flock, a normal layers' mash may be fed. Breeders' mash should, however, be re-introduced four weeks prior to hatching eggs being required.

Feeding males

Breeding males often fail to eat sufficient food to maintain condition and this failure may result in loss of fertility. A few (one per ten males) tubular feeders may be suspended at a height which allows the males to feed, but is too high for the females. Half the feeders should be filled with normal breeders' mash and the remainder with whole wheat or kibbled maize.

Specimen formulation for broiler-breeder diet

Ingredient		*Broiler breeder*
Maize	%	24·90
Wheat	%	40·00
Middlings	%	11·30
Herring meal	%	1·20
Soya-bean meal	%	7·10
Meat and bone meal	%	4·60
Grass meal	%	2·50
Dried distillers' solubles	%	2·50
Methionine	%	0·044
Limestone	%	5·40
Salt and vitamin supplement	%	0·50

This yields:

ME	kcal/kg	2,750
Crude protein	%	15·90
Lysine	%	0·72
Methionine	%	0·32
Methionine plus Cystine	%	0·59
Oil	%	2·80
Fibre	%	3·80
Calcium	%	2·70
Available Phosphorus	%	0·42

Ration formulation for ducks and geese

SUMMARY FEEDING STANDARDS FOR DUCKS HOUSED INTENSIVELY

0–4 weeks	*Quoted range in formulation*
Type of food: Duckling starter	
Form of food: Crumbs or wet mash	
Crude protein content	18%–22%

	Quoted range in formulation
0–4 weeks	
Metabolizable energy:	
kcal/kg	2,750–3,300
kcal/lb	1,250–1,500
4–8 weeks	
Type of food: Duck finisher	
Form of food: 3·2 mm pellets or wet mash	
Crude protein content	16%–18%
Metabolizable energy:	
kcal/kg	2,750–3,300
kcal/lb	1,250–1,500
8 weeks onwards (breeding stock)	
Type of food: Duck growers	
Form of food: 4·7 mm pellets or wet mash	
Crude protein content	14%–16%
Metabolizable energy:	
kcal/kg	2,640–3,080
kcal/lb	1,200–1,400
Breeders	
Type of food: Duck breeders	
Form of food: 4·7-mm pellets or wet mash	
Crude protein content	16%–20%
Metabolizable energy:	
kcal/kg	2,400–2,900
kcal/lb	1,090–1,320

FORMULATION NOTES

Assumptions concerning performance of production stock

Strain of duckling has a significant effect upon the rate of growth and food conversion efficiency in a table-duck enterprise.

The strain selected should grow rapidly to meet the market requirements for the older the ducklings are when killed the less satisfactory will be their food conversion. This is well demonstrated in the figures in Table 32.

Table 32 Performance of five strains of ducklings – killing age 49 days (*Cherry Valley Farms 1975*)

Liveweight (kg) at slaughter mixed sexes	3·1	3·05	2·96	2·72	1·96
Food conversion kg/kg liveweight	2·85	2·86	2·98	2·91	3·1

The figures cover the normal range of performance as the figures in Table 33 calculated for oven-ready weight show, though these fall a long way behind the record performance levels of small pens of birds on performance test (Table 34).

Table 33 Average results obtained from one farm producing several million oven-ready ducklings per year (*Cherry Valley Farms 1975*)

Oven-ready weight	2·136 kg
Kg food per kg of oven-ready duckling	3·9:1
Killing-out percentage	73%

Table 34 The world record. Killing age 56 days (*Cherry Valley Farms 1975*)

Liveweight	*Grams*	*Pounds*	*Food conversion*	*Mortality*
Males	4·307	9·50	3·10	1/80th
Females	3·789	8·37	3·26	—
Average	4·053	8·935	3·18	1/60th

Assumptions concerning the performance of breeding stock

Table 35 Results of breeding stock (*Cherry Valley Farms 1975*)

Length of season	40 weeks
Number of hatching eggs	176–205
Hatch of all eggs	75%–82%
Mortality	5%

Table 35 (*continued*)

Food/day-old duckling produced (i.e. food of breeding male and female divided by the number of ducklings)	618 grams
Annual output of ducklings per m²	83·95 kg

Assumptions concerning management standards (*Poultry World 1975*)

Stocking density

System	*Ducks/m²*	*Sq ft/Duck*
Range	0·03–0·04	250–300
Straw yards	1–2	4–8
Verandah	4·6	1·5–2·5

Food consumption per 100

Age	*Kg*	*Cwt*
0–4 weeks	205	5
5–8 weeks	760–1,015	15–20

Feeding space per 100

Age	*Length of trough*	
	Metres	*Feet*
0–4 weeks	3	10
5–8 weeks	6	20
8 weeks onwards	9	30

Targets 3·2 kg (7 lb) liveweight at eight weeks with a food conversion ratio of 3·5:1.

Breeds and adult weights

Breed	*Kg*	*Lb*
Indian Runner	1·59	3·5
Campbell	2·04	4·5
Muscovy	2·49	5·5
Orpington	2·60	5·75
Pennine	3·17	7·0
Pekin	3·63	8·0
Aylesbury	3·85	8·5

Vitamin E

Ducklings are very sensitive to deficiency of vitamin E and selenium, which may result in degeneration of the skeletal and heart muscles.

Wet mash

Wet mash may be used as an alternative to crumbs and pellets. The ducklings should be given as much as they can consume in thirty minutes three times a day. The wet mash should be made up in a suitably crumbly state and not as a doughy mass. The use of wheat offals in the food mixture will help to produce a suitable wet mash.

Alternative foods

It is not always possible to purchase special ready-made foods for ducklings and, in such cases, chick starter crumbs may be used followed by pellets.

Insoluble grit

A little chick-sized insoluble grit should be sprinkled on the food during the first few days and after this access allowed to a hopper of coarse-size grit. If a constant supply is not given a small amount should be provided with the day's food at two-weekly intervals.

Antibiotics

There is little evidence to suggest that the use of antibiotics is of value in the food of ducklings for they are normally very free from diseases.

SUMMARY FEEDING STANDARDS FOR GEESE

Geese housed intensively and in yards

Information regarding the nutritional requirements of geese is very limited and specially made rations are rarely available. It is common practice to use formulations designed for ducks or chickens instead.

	Quoted range in formulation
0–4 weeks	
Type of food: Starter	
Form of food: Crumbs or wet mash	
Crude protein content	20%–24%
Metabolizable energy:	
kcal/kg	2,750–3,300
kcal/lb	1,250–1,500
4–10 weeks	
Type of food: Growers No. 1	
Form of food: 3·2 mm pellets or wet mash	
Crude protein content	20%–24%
Metabolizable energy:	
kcal/kg	2,750–3,300
kcal/lb	1,250–1,500
10 weeks onwards	
Type of food: Growers No. 2	
Form of food: 4·7 mm pellets or wet mash	
Crude protein content	16%–18%
Metabolizable energy:	
kcal/kg	2,420–2,640
kcal/lb	1,100–1,200

Geese on range

Geese require about four times the amount of space normally allowed for chickens, so that it is very convenient to rear them on range. Goslings may be reared on grass range from three days of age according to the climatic conditions, and from then onwards the range system is entirely suitable.

The earlier the goslings form the habit of grazing the better. Once the goslings are fully used to the open air (this is usually by the end of the first week), the feeding of balanced rations should gradually be reduced in order to encourage the grazing. By the end of the fourth week the sole diet may consist of green food. The total amount of starter food used may be in the region of 500 g

(11 lb) per gosling. If the quality of the range is poor, additional food will be required.

In a good summer supplementary feeding need not be recommended until the autumn when the feeding value of the grass range falls. The food used during the fattening period, i.e. four weeks, may amount to about 62 kg (28 lb) of whole grain per goose. The geese should be allowed to eat to appetite.

Goslings may be fed their requirements of balanced foods as a wet mash. The wet mash should be provided three times daily and the goslings allowed to eat to appetite for a period of half an hour.

Breeding birds kept on range can be stocked at the rate of 12–24/hectare (5–10/acre). Supplementary feeding for hatching-egg production should commence in February. The ration will have the following characteristics:

Type of food:	Breeders	
Form of food:	4·7 mm pellets or wet mash	
Quantity	110 g (4 oz) per goose per day plus 110 g grain	
Crude protein content		16%–20%
Metabolizable energy:		
kcal/kg		2,400–2,900
kcal/lb		1,090–1,320

Table 36 Breeds of geese and liveweights (*Poultry World 1975*)

Breed	*Liveweight* *kg*	*lb*
Embden	7·3–10·9	16–25
Toulouse	6·3–9·1	14–20
Brecon Buff	7·3–8·2	16–18
English	5·4–9·1	12–20
Chinese	3·6–5·4	8–12
Roman	4·5–6·3	10–14

10-week geese

Large rapidly growing geese may be used for the so-called 'green goose' trade when they are confined and fed exclusively on generous quantities of balanced rations. Such geese must be ready for killing at about 9–10 weeks of age, since rate of growth declines rapidly after 8 weeks of age.

Liveweight at 9–10 weeks should average 4–4·5 kg (9–10 lb) with a food conversion ratio of 2·5–3·5:1.

Ration formulation for turkeys

SUMMARY FEEDING STANDARDS FOR INTENSIVELY KEPT TURKEYS

	Quoted range of formulation
0–4 weeks	
Type of food: Turkey pre-starter	
Form of food: Mash or No. 1 crumbs	
Crude protein content	28%–32%
Metabolizable energy:	
kcal/kg	2,640–3,080
kcal/lb	1,200–1,400
4–8 weeks	
Type of food: Turkey starter	
Form of food: Mash or crumbs	
Crude protein content	22%–26%
Metabolizable energy:	
kcal/kg	2,640–3,080
kcal/lb	1,200–1,400
8–11 weeks (females); *8–12 weeks* (males)	
Type of food: Growers No. 1	
Form of food: Mash or 3·2 mm pellets	
Crude protein content	20%–24%
Metabolizable energy:	
kcal/kg	2,640–3,080
kcal/lb	1,200–1,400
11–14 weeks (females); *12–16 weeks* (males)	
Type of food: Growers No. 2	
Form of food: Mash or 4·7 mm pellets	
Crude protein content	18%–22%
Metabolizable energy:	
kcal/kg	2,640–3,080
kcal/lb	1,200–1,400

14–17 weeks (females); *16–20 weeks* (males)	*Quoted range of formulation*
Type of food: Growers No. 3	
Form of food: Mash or 4·7 mm pellets	
Crude protein content	14%–18%
Metabolizable energy:	
kcal/kg	2,640–3,080
kcal/lb	1,200–1,400

17–20 weeks (females); *20–24 weeks* (males)

Type of food: Pre-breeders	
Form of food: Mash	
Crude protein content	14%–15%
Metabolizable energy:	
kcal/kg	2,640–3,080
kcal/lb	1,200–1,400

Adult breeders

Type of food: Breeders	
Form of food: Mash	
Crude protein content	14%–15%
Metabolizable energy:	
kcal/kg	2,640–3,080
kcal/lb	1,200–1,400

FORMULATION NOTES

Assumptions concerning performance of turkeys (*Poultry World 1975*)

Table 37

Age in weeks	*Liveweight Stags kg*	*lb*	*Hens kg*	*lb*	*Cumulative food conversion Stags*	*Hens*
8	3·2	7	3·35	5·17	1·65	1·81
12	5·89	12·9	4·53	9·96	1·97	2·20
14	7·37	16·2	5·57	12·2	2·18	2·38
16	8·84	19·4	6·46	14·2	2·37	2·60
18	10·25	22·5	7·25	15·9	2·55	2·82
20	11·61	25·5	7·93	17·4	2·75	3·08
24	14·15	31·1	9·9	21·8	3·29	3·70

Turkey breeders

Potential	70 saleable poults per hen
Average	50–60 poults per hen

Assumptions concerning management standards

Stocking density

	Liveweight	*Birds per m²*	*Sq ft per bird*
0–12 weeks			
Males	5·89	5·0	2·16
Females	4·53	6·5	1·66
12–24 weeks			
Males	14·15	2·0	5·2
Females	9·9	3·0	3·6
Breeders	9·9	3·0	3·6

Food hopper space

First five days	One clean tray per 100 poults
Growers and adults	Six round troughs, or pans, per 100 turkeys

Water supply Ten round automatic troughs per 1,000 turkeys.

Ventilation 11 m³ minimum, 90 m³ maximum of air per hour per kilo of food consumed daily.

Nutrient levels

The nutrient requirements of male and female turkeys differ and separate rearing may lead to savings in food consumed.

Amino acids

Turkey rations need supplementation with methionine and sometimes with lysine. Lysine is of particular importance for it is concerned with proper feathering. Bronze poults fed on a ration deficient in lysine show a characteristic white barring of the primary and secondary feathers of the wings.

Folic acid

The level of folic acid in breeders' formulations should be checked.

Coccidiostat

A coccidiostat will generally be required in the starter food if the turkeys are reared on litter and may be required in the diet of turkeys reared on other systems.

Blackhead

A drug to prevent this disease should be included, following the advice of a veterinarian, if the turkeys are reared on the floor.

Growth promotant

A growth promotant may be included in the rations for turkeys being reared for table purposes but not in the ration of potential breeding stock.

Restricted feeding

Restricted feeding will often be required during the rearing stages of breeding turkeys. The advice of the breeder of the stock should be followed.

FEEDING GREEN FOOD

Turkeys readily eat fresh green food and if given the opportunity will consume large quantities of it which may be used to supply up to 10 per cent of the dry matter of the total diet. Green food contains a considerable proportion of water so the actual proportion that it contributes to the daily food intake on a wet basis may be as high as 40 per cent.

Fresh lucerne is a first-class green fodder for turkeys of all ages. Grown properly, it will yield up to four cuts a year, for six years in succession. The yield may be greatly improved by the liberal use of turkey manure on the field in the winter. Thousand-head kale is another excellent means of supplying green food.

Where, in temperate climates, an attempt is made to provide an all-the-year-round supply of green food for turkeys, lucerne may be used for six months followed by kale for three to four months with curly kale to fill the gap until the spring regrowth of lucerne. Adult turkeys may be made to graze their own green food with a policy of strip grazing.

Although turkeys will consume large quantities of green food it is not always desirable to allow them an unrestricted supply. During the early season when the fodder is of high quality, the amount that they consume may be almost unlimited. At this stage it is extremely palatable, has a very high nutrient value and contains an appreciable quantity of protein and pro-vitamin A. Later in the season, however, its nutritional value is considerably lower, and its very high fibre content may lead to digestive troubles (see Chapter One, pages 24–26) and to a reduced rate of growth.

When turkeys are given access to good-quality range, the birds should be encouraged to forage throughout the day by arranging to supply the bulk of their 'balancer' food an hour or two before dusk. A suitable ration may be fed together with whole grain. The 'balancer' food is best supplied in the form of pellets.

FREE-CHOICE FEEDING

One method of feeding growing turkeys after eight weeks of age is to allow access to a protein/mineral/vitamin concentrate in one hopper and to whole wheat, oats or kibbled maize in another. The term 'free choice' is often given to this method of feeding. Another variant is to allow the turkeys a choice of three hoppers: concentrate in one, wheat or maize in another, and whole oats in the third.

Growing turkeys will readily consume whole cereal grains, particularly whole oats, and providing that the total diet is correctly balanced there is no limit to the amount of whole grain that may be fed. Allowing the turkey free choice of grain and concentrate may seem a little empirical but turkeys make a good job of balancing things up. Cereal grains are deficient in proteins, minerals and vitamins, and these must be made good by a suitably balanced concentrate, which is usually referred to as the grain booster or grain balancer.

In practice, the free-choice method of feeding appears to be only slightly less efficient that the provision of a carefully balanced diet, when considered from the point of view of food conversion. However, in terms of food cost per pound of meat produced, the free-choice method often shows up very well, particularly when cereals are available on the farm or can be purchased at a very low cost.

An interesting experiment was conducted in Georgia where the turkeys had free choice of concentrate, oats and maize throughout the period from 8–24 weeks. The free-choice Broad Breasted Bronze birds were only slightly less efficient in food conversion than the control-fed group. The grain balancer used in this trial was a standard ration with defluorinated phosphate increased from 1·5 to 3 per cent at the expense of maize.

WET-MASH FEEDING

The wet-mash method of feeding turkeys is not generally recommended. If it is to be used, three feeds daily of wet mash should be given during the first week, followed by two feeds daily.

As an alternative to the use of wet mash only, one method of feeding is to combine wet mash with the use of whole cereal grains. A small feed of up to 15 g (0·5 oz) of whole grain may be given first thing in the morning, a full feed of wet mash at midday, and a full feed of grain in the evening. The wet mash should be specially formulated to balance the deficiencies of the grain.

The presence of up to 30 per cent wheat offals in the formula will result in a wet mash of better consistency. High levels of ground wheat may result in a doughy and unpalatable wet mash.

Specimen formulations for turkey diets

Ingredient		*Turkey starter*	*Turkey grower*
Maize	%	16·00	8·30
Wheat	%	37·50	50·00
Barley	%	—	8·30
Herring meal	%	10·00	5·00
Soya-bean meal	%	27·40	19·40
Meat and bone meal	%	5·00	5·00
Dried distillers solubles	%	2·50	2·50
Methionine	%	0·45	0·45

Ingredient		*Turkey starter*	*Turkey grower*
Limestone	%	0·50	0·60
Di-calcium phosphate	%	0·60	0·50
Salt and vitamin supplement	%	0·40	0·40
Coccidiostat		+	+
Anti-blackhead		+	+

These yield:

ME	kcal/kg	2,860	2,860
Crude protein	%	28·10	22·50
Lysine	%	1·65	1·20
Methionine	%	0·57	0·45
Methionine plus Cystine	%	1·00	0·81
Oil	%	2·90	2·50
Fibre	%	3·70	3·70
Calcium	%	1·30	1·10
Available phosphorus	%	0·75	0·60

CHAPTER NINE

Feeding methods and appliances

Forms in which food may be offered

INTRODUCTION

Previous chapters have made it clear that the food offered to domestic poultry is often in the form of a complex mixture of ingredients, blended together to include, in any sample, an appropriate balance of nutrients which will meet the total requirements of the birds to which it is being fed. The production of such rations is the logical outcome to the scientific study of the nutrition of the fowl, in which an analysis of its metabolic needs has led to the synthesis of a ration mixture which comes close to a guarantee to meet those needs. The problem, from the nutritionist's point of view, is solved.

It is important, however, to remember that there are further consequences of achieving this solution, which remain important in its practical application. These include the overall costs of the process, its efficiency in providing a uniformly mixed diet and the ease with which the food may, on the one hand, be handled by the farmer and, on the other, be eaten and digested by the birds.

Chapter Seven set down the number of times that food ingredients have to be handled or moved in the mill before a completed ration can be sent out to a farm. Despite the savings afforded by mechanization, every one of the processes described accounts for an increase in the cost of the diet. When one remembers, in addition, all the overhead charges of the mill's operation, it is obvious that the preparation of the nutritionist's recommendations into a blended diet is an expensive process. It is likely that the advantages of this carefully formulated ration will justify the expense of preparing it, but there is little wonder that the farmer should be interested in saving some of the costs and in exploring the possibilities of unprocessed ingredients in their natural state.

The most obvious example would be that of the cereal grains which can be fed whole or kibbled to chickens, when they are not ground into the fine particles in which they can be more readily mixed with other ingredients. It is important, therefore, to know if the feeding of whole grains (and of various other nutrients) is a satisfactory way of presenting the bird's diet, since it clearly would be cheaper.

From another point of view it is necessary to know whether the finely divided meal, required for achieving thorough mixing, is the most desirable material for subsequent handling and feeding. The waste in the mill, caused by the dust of over-finely ground particles, is likely to continue after mixing, whereas the pressing of the meal into pellets and crumbs not only helps to control dustiness but also increases food intake by most classes of fowl. What then is the impact of these factors on the decision concerning the way in which food shall be given to birds?

Feeding compounded foods

If we work first from the assumption that the nutritionist's work has indeed been a necessary part of the process of feeding chicken, then the first form of food to consider is the newly mixed meal.

Dry mash

Supplying dry mash is generally the cheapest method of feeding compounded rations when considered in terms of price per tonne because, of course, the pelleting cost has been avoided. Generally speaking, mash should be made available at all times, i.e. fed *ad libitum*. It is also essential for the mash to be freshly mixed. Stale food may result in reduced nutritional value and/or reduced intake, which will be reflected in rate of growth or production. Musty foods should never be fed to young stock. The use of freshly mixed mash is particularly important in hot climates where food may rapidly become dry and dusty, and if stored for too long it is likely to turn rancid.

Special care should be given to particle size, the 'grist', of the mash to ensure maximum palatability. Wheat in particular should be coarsely ground because finely ground wheat meal contains a high proportion of flour which tends to paste and stick to the inside of the chicken's beak, resulting in impaction of the beak or the tongue. This will be followed by a fall in food intake and consequent reduced rate of growth.

Again very finely ground mash leads to waste since if it is spilled for any reason it is difficult to gather up again from most of the surfaces on which it will fall, without contamination. Even the birds themselves will be unable to pick it all up, in marked contrast to the way in which they can recover spilled grain from almost any surface.

On the other hand, mash is easy to store and handle, though some individual ingredients are at risk and, subject to a small percentage of errors made in mixing, it is always likely to be a fully balanced diet.

Wet mash

The very dryness of dry mash sometimes reduces food intake below the desired level. This is particularly true if water intake is low either because the water supply is some distance away or because it has become dirty or unacceptable for some other reason. In these circumstances intake can nearly always be brought back to normal by adding water to the mash.

If it is decided to adopt a wet-mash feeding system, three feeds daily will be necessary during the first week of the bird's life, after which two feeds a day will be satisfactory. Birds should be allowed to consume as much food as they desire at each meal and the amount to be prepared can usually be judged after a little experience.

The first meal may be given early in the morning, and sufficient wet mash should be supplied so that the birds clear it up within thirty to sixty minutes before the second feed is due in the afternoon. If it is left in the hoppers, wet mash will rapidly become sour and mouldy and, for this reason, it is important to clean out and scrub the hoppers at least once a week.

Ample hopper space is required with wet-mash feeding; sufficient to allow all the birds to feed at the same time. To ensure that the birds really are consuming sufficient food they may be handled in the evening to check that their crops are full.

Wet mash should be made into a suitably crumbly state so that when a handful is picked up and released, it immediately falls to pieces. If the mash becomes soggy and dough-like on wetting, this should be considered as quite unsuitable for young birds.

Clearly, there are difficulties to be faced with wet-mash feeding and its usefulness is, therefore, limited. It is most suitable for making use of home-produced foods by the small-scale farmer. It may, for example, be convenient to use a large proportion of potatoes in the diet. For success, these must be cooked, virtually to a porridge, which mixed with meal gives wet mash.

Similarly wet-mash feeding is a convenient way of utilizing some industrial by-products such as biscuit meal and condensed fish solubles.

The consistency of the wet mash depends to a great extent upon the constituents and some ingredients may be used specifically for this purpose. Thus it is advisable to include a minimum of 20 per cent of millers' offals in the mixture. A ration containing 20 per cent middlings and 20 per cent bran can be made up into a very well-textured wet mash.

By contrast, the inclusion of ground wheat in rations intended to be used as wet mash should be limited to 20 per cent and the wheat should be coarsely ground, otherwise the flour may result in a sticky dough.

The difficulties, however, are not solely physical. If unusual foods are being offered to the birds, special care must be taken to ensure that the meal used to complete the mash gives a proper nutritional balance with potatoes or fish solubles or whatever less expensive product may be locally available. If a continuing supply of such an ingredient is available, it is sensible to have it analysed so that a balancer meal can be specially prepared.

A combination of wet mash with dry feeding may be used and will give good results. When wet mash and pellets are employed, a few pellets may be given first thing in the morning, the wet mash fed at midday, and a full feed of pellets given some time during the afternoon. Where the dry portion of the diet consists of dry mash it is wise to provide an *ad-libitum* supply. The wet-mash portion of the diet may then be given in the morning or at midday. Separate hoppers are generally advisable but, if necessary, wet mash may be placed on top of the dry mash.

Crumbs and pellets

Crumbs followed by pellets is the system of choice for broiler and capon production, where the aim is to achieve the maximum food intake together with the best rate of growth and food conversion ratios.

The crumbs and pellets should be hard. Soft pellets and crumbs tend to break up, and the mash so formed collects in the base of the food containers, which may them become blocked.

Crumbs are made from 4·7 mm ($\frac{3}{16}$ in) pellets. Growers' pellets are of 3·2 mm ($\frac{1}{8}$ in) size and adult pellets are of 4·7 mm size.

Crumbs and pellets are very easy to handle and convenient to use. Once chickens have become used to them, a change to mash feeding should only be made gradually in order to allow the birds to become accustomed again to mash.

The employment of crumbs and pellets is useful on occasions because the process of pelleting may render certain foods more palatable, thus occasionally allowing the use of ingredients that would not be suitable for a mash-feeding system.

Birds normally show an increase in intake of 6 to 8 per cent when they are fed pellets. This may be wasteful if it means that they now eat more than will provide for their normal requirements. Pellets may also lead to another form of waste in that battery-caged birds sometimes will flick large quantities of pellets with their beaks on to the floor of the house or into the dropping trays of their cages. Losses as high as 8 per cent may occur in this way.

FEEDING UNPREPARED FOODS

Feeding grain

Ducks, chickens and turkeys readily consume whole or cut grain and, where supplies are available, perhaps home-grown grain feeding can be an inexpensive way of providing part of the diet.

Cereals are deficient in proteins, minerals and vitamins and it is essential to make good these deficiencies. The mash or pellets used to supplement the grain must contain additional nutrients to ensure that the requirements of the birds are met. Suitable mixtures known as grain balancers are available.

A suitable feeding programme is as follows. The grain balancer mash is available at all times and the grain is fed first, in the morning, at the rate of say 10 g per layer; a full feed of grain is then given in the evening, about an hour before dusk, allowing 40 g per bird.

The alternative is to mix up the balancer as a wet mash and to provide this at midday, allowing the birds to eat to appetite for half an hour. Again instead of wet mash, the birds could be given a midday feed of grain balancer pellets at the rate of 40 to 50 g per bird daily.

Whenever grain is fed, the birds should be provided with a supply of insoluble grit (granite or flint grit) of suitable size, to ensure that the grains do not pass through the body unutilized.

Green food

Although poultry of all types will readily consume fresh green food, it should not be considered an absolutely essential part of the diet. First-quality dried grass or lucerne may be used instead and is frequently included in a prepared diet at a level of 2·5 to 5 per cent.

The decision whether to use fresh or dried green food will vary according to the particular circumstances of the farm. The large farm that already has a grass-drying plant may find it more convenient or economical to feed dried grass rather than to supply birds with fresh green food.

If given the opportunity, fowls will consume large quantities of greenstuffs. Some farmers claim that if turkeys especially are given access to really first-class green food it is possible to make a considerable reduction in feeding costs. For chickens good-quality green food is also useful and may be used to supply up to

10 per cent of the total diet on a dry-matter basis. Green fodders contain perhaps 80 per cent or more of water, so the actual portion of the daily food intake on a wet basis may be equal to 40 per cent of the diet.

This, of course, is the reason that makes range management of poultry a physical possibility. Particularly with small flocks, it is possible to keep the birds out of doors in grass pens and to expect that, whilst the grass is good during the spring and summer seasons, their need for prepared foods will be reduced to approximately 50 per cent of that which they would eat indoors. To be operated satisfactorily, this system demands a relatively low stocking-rate of 250 per hectare (100 per acre) and the regular alternation of the grazing areas of the birds. It is expensive in land use and requires plenty of good skilful labour. Many people say that birds on this type of system produce the best eggs for eating.

Greenstuffs can be fed in intensive poultry houses. In deep-litter pens, bunches of lucerne (*Medicago sativa*) or fresh-cut brassicas may be suspended for the birds to peck at. In cage units chopped lucerne spread along the trough is nearly always eaten greedily by the birds. There are dangers, however, for the greenstuff is not likely to be high in energy content and its consumption may depress the birds' intake of essential nutrients below the desirable level.

Free-choice feeding

The systems of feeding unprepared foods would seem to require extra care in the provision of balancing rations, especially if the birds have free access to grass or other green fodders where there is little chance of controlling their intake of the latter foods. In fact, there is some evidence to suggest that if a mean intake figure for the unprepared food can be assumed, then a balancer can be prepared on that assumption and be offered for *ad libitum* consumption by the birds. In other words the stock (adult chickens anyway) have the ability to establish a balanced intake from the two sorts of food if they can pick from these at will. Experiments conducted at Edinburgh University suggest that layers, regulating their own intake in this way, have a comparable nutrient intake with that of the controls and do not show any fall in performance. This finding is very interesting since it might lead to substantial savings in the cost of food preparation, but it is not yet possible to generalize it for any great range of unprepared feedstuffs each of which might form the basis of a daily ration. Nor is it known under what circumstances gross overeating might occur.

Feeding systems

Free-choice and dry-mash feeding are both commonly operated on an *ad-libitum* basis. That is to say the food is always available and the bird can eat at will. Laying birds, in particular, have been thought to be able to regulate their intake closely to their requirements. If anything, it has been felt that the problem might be to get enough food into the birds rather than that there should be serious danger of costly overeating. As long as this supposition might be approximately true and until the prices of food supplies rose so rapidly over recent years, it was convenient to accept it because *ad-libitum* feeding is easy to manage. Supplies can be set out for several days at a time and no monitoring or supervision of the bird's daily intake is required. Systems of mechanical delivery of food to the birds are relatively easy to devise for this situation.

The conditions noted above no longer hold. The causes of variation in food intake have been more closely observed in the last few years and are discussed in Chapter Eight, pages 145–151. There are many circumstances in which birds overconsume both energy and the total ration, a fact which must lead to overexpenditure on food. In addition, now that food prices are so high any saving that can be made in intake of food (which does not affect performance) must be worth investigating. For this reason, systems of restricted feeding now command very great interest.

RESTRICTED FEEDING

Allowing birds an unlimited supply of food may, with some strains, result in heavy overconsumption. The birds are liable to consume too much energy and this will be converted into body fat which in turn leads to a higher maintenance requirement. In addition birds that become too fat tend to give reduced egg production, and infertility and mortality may be increased.

Overconsumption is particularly liable to occur with strains of chickens and turkeys selected for meat qualities. Restricted feeding programmes are therefore recommended for potential broiler breeders and with some strains of turkeys. A restricted feeding programme may sometimes be recommended for some strains of laying chickens, although in this case considerable care in the application of the programme is required. The degree of food restriction suitable for layers would appear to be much more limited than for broilers, and over-restriction can have adverse effects.

Some other aspects of savings from restricted feeding are important. If the poultry house is efficiently constructed to give a high level of heat conservation, restricted feeding may be practised to take advantage of the fact.

Again restricted feeding may be used as a part of a programme to delay sexual maturity. In this case it will be coupled with a suitable lighting pattern that has the same objective.

It is known that there is a correlation between bodyweight and maturity. With most strains of broilers a restricted feeding programme is essential during the growing stage to retard liveweight gain and to delay sexual maturity thus preventing the birds from becoming too fat and allowing them the best chance of economic production.

In the majority of cases the recommended feeding programme of the breeder should be followed carefully and only departed from for very good reasons. Over-restriction of feed is also harmful and can result in very late maturity and unsatisfactory production.

Restricted feeding programmes are often started with replacement broiler breeders at about eight weeks of age and continued throughout their laying life.

Percentage restriction

Several ways have been investigated to limit the bird's food intake. Birds may be given time-limited access to the food troughs or the troughs may contain a reduced amount of food. It was quickly recognized that restriction must not be on an empirical basis since over-restriction might be as harmful as overconsumption. What was missing to start with was an appropriate base line from which to calculate the restriction.

In percentage restriction the daily food intake is limited to a certain proportion of the known intake of similar birds fed *ad libitum*. In the case of broiler breeders the intake may be restricted to 70 per cent of the normal intake. Successful application of the system includes the regular weighing of the birds to ensure that they reach the correct weight for their age. In this way the amount of food provided can be increased or decreased in order to maintain the liveweight gain at the required level. The breeders of most strains of broilers have prepared detailed programmes for restricted feeding and these should be followed. These programmes have translated what was originally percentage restriction to a table of recommended absolute daily intake levels varying with age. Contemporaneous monitoring against birds fed *ad libitum* is not required.

Similar restricted feeding of laying birds during the rearing period is not yet widespread and cannot be recommended generally. With adult layers some success has been claimed for limiting the food intake to 95 per cent of the *ad-libitum* level from about forty weeks of age. With this system it is recommended that a small group of birds should be kept on full feeding so that the normal intake can be accurately assessed. Food intake tends to fall during periods of hot weather and according to the rate of production. It is, therefore, necessary to keep a regular check on the consumption by the control group.

The system makes heavy demands on labour since satisfactory mechanical means of adjusting food intake accurately in this way have been slow to appear.

Limited access – 'skip-a-day' feeding

Even with continuous access to their food birds may show distinct peak times of eating, early in the morning and just before dark. By exaggerating feeding activity at these times they can eat enough for a twenty-four-hour period. To be successful, therefore, limited access must be continued for as long as a day. In this method the birds are supplied with their normal food, *ad libitum*, for six days a week. On the seventh day the birds are given 800 g (2 lb) of grain only per 100 birds. The skip-a-day programme is the most simple method of restricted feeding but really what it chiefly saves is labour on the seventh day. As far as food intake is concerned the birds tend to eat more on the other six days.

Varying nutrient density

Another approach to the restriction of nutrient intake has been that of diluting the nutrient density of the ration. This was first attempted by mixing into the ration a substance, preferably a cheap one, which was relatively inert nutritionally. Cereal offals and straw have received the most attention but even sand has been used experimentally. The first effect of such dilution is to make the birds eat more of the food in order to maintain their energy intake. Unless the dilution is very severe modern layers are successful in this compensatory eating. When, as has happened, the diluting materials become expensive, the attempt at economy is even more frustrated.

If energy dilution is unsuccessful, will the reduction of the protein content of the diet serve any better purpose? There is some evidence that perhaps the dilution of certain amino acids might do so. Thus in South African experiments, birds fed between the ages of 6–19 weeks on diets containing approximately 60 per cent of normal levels of lysine laid slightly fewer eggs than controls but at lower cost in food overall. The birds ate less food during the period of restriction but tended to compensate later when they reverted to a normal diet. More research is required before practical recommendations can be made.

However, one aspect of protein restriction or saving has received practical support and that is the practice of 'phase feeding', which is based on the idea that laying birds have different nutritional needs at different 'phases' of their productive life. The most common example is the practice of feeding mixtures of standard nutritional content to laying birds up to say 40–42 weeks of age and then reducing the protein level and using a cheaper food for the remainder of the laying period. The possible dangers of this practice have already been discussed in Chapter Eight, pages 163–165.

Close rationing

The two previous sections have demonstrated that poultrymen are interested in restricting the food intake of their birds and also that they are liable to be thwarted by the high labour costs of precise rationing. Systems which seek to avoid high labour use have only very limited success. Some hope, however, still exists.

A recent development in mechanical feeding equipment (see page 146) allows considerable precision in rationing birds housed in battery cages. The mechanism delivers food to the trough in front of one cage (per row) at a time. It delivers a weighed allocation equal to perhaps 1/12 or 1/20 of the birds' calculated intake and this amount forms a thin layer in the bottom of the trough. The bird (or birds) in the cage quickly eat this 'snack' and are very ready for the next one delivered thirty minutes later.

The idea depends on the concept of an appropriate daily food intake for the birds being fed and on the knowledge that the ration has been formulated in accordance with that intake (just as the A.R.C. Tables of Requirements use the same assumption as a base). Once the assumption is made and is sound, it only remains to arrange a system whereby precisely this amount of food can be offered to each bird.

The limitation of the total food offered per day to this standard prevents wastage from overconsumption and inappropriate usage of the diet. The division of the amount into small portions ensures that physical wastage is minimal, there not being enough in the trough to be spilled or flicked out. The trough stays clean and less food is spoiled by overflow from the drinkers. Savings in expenditure on food of 20 per cent or more have been claimed for this approach, which figure serves to indicate the amount of wastage that may occur in less-well-managed systems.

Feeding appliances

INTRODUCTION

Food is nearly always offered to the bird in some form of trough – usually either long and thin in shape or round. The function of the trough is twofold; it must give the bird easy access to the food but it must also preserve the food from being spilled and from being contaminated by the bird's droppings or by litter and dust from the floor. These requirements are to some degree contradictory since a trough wall which is high enough to minimize wastage may also be too high to allow young birds to feed in comfort.

Simple troughs become more complicated when they are attached to some form of closed food container. In this form the appliance is called a hopper and the most common form is the round tubular feeder. The hopper needs to be filled at intervals, according to its size and capacity, but its function is to ensure that food is in the trough and available to the birds at all times.

The most popular tubular feeders are made either of galvanized iron or plastic. They are made to hold 12 to 15 kg (26 to 33 lb) of food, and will be 38 to 45 cm (15 to 18 in) in diameter. The same hoppers may be used for both growing and adult birds.

Various criteria concerning food troughs and hoppers must always be borne in mind. The main ones are the following.

Trough space

It is essential to provide ample feeding space so that all the birds can feed at one time. Insufficient feeding space may result in uneven growth and poor feathering. This is especially true if rationed amounts of food are being supplied. It is less critical if food is available *ad libitum* for then birds will learn to approach the food trough at different times if it is not big enough to accommodate all at once, 2·5 cm (1 in) per bird is sufficient for young chicks but adults will require 10 cm (4 in) each. The particular virtue of round or tubular feeders is that, as the birds stand in a circle round them, the width of troughing they require is limited to the space needed for their heads. Their bodies are accommodated in a wider circumference, further from the centre of the hopper.

Floor troughs

Floor troughs are the traditional form of feeder but though they are still used they are not generally recommended. Their main disadvantage is that they occupy valuable floor space, restricting both movement of the birds and the stockman's ability to manage the litter. Two or three different sizes of trough are necessary as the birds grow and with all of them it is difficult to control wastage of food (see Figs. 7 to 10).

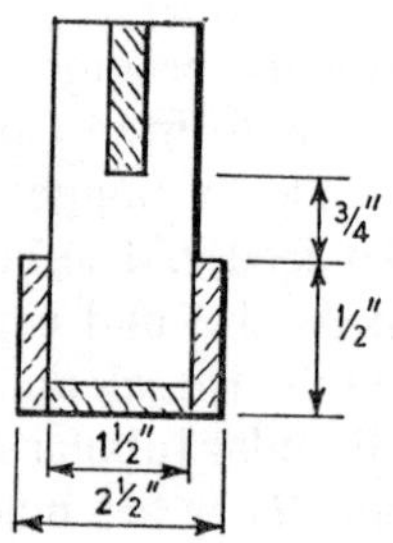

Fig. 7 Cross-section of feed hopper for use during the first 3 to 6 weeks

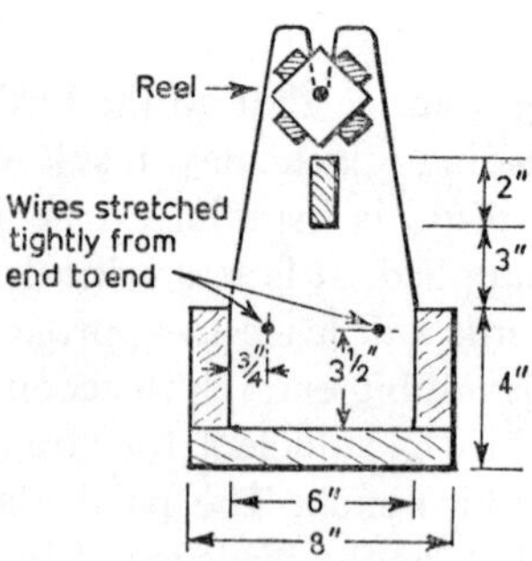

Fig. 8 Hopper for use 6th to 12th week

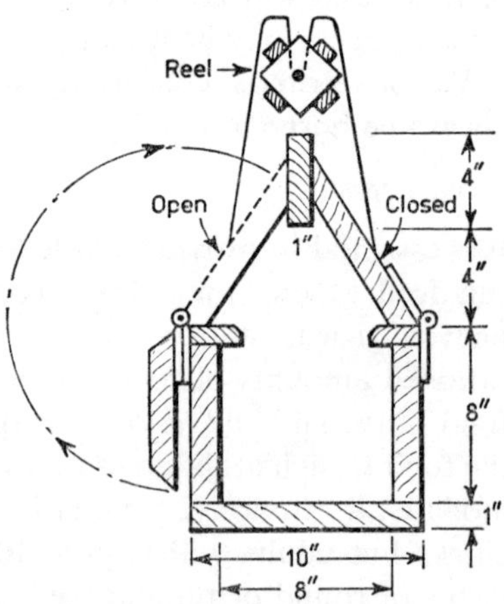

Fig. 9 Hopper for adults

Troughs should be raised off the floor on bricks as the birds grow, especially turkeys, in order to keep litter out of the troughs and to encourage a better food intake. They should also be moved from one part of the house to another at intervals, thus allowing better distribution of droppings, and preventing the appearance of wet and dirty spots in the litter.

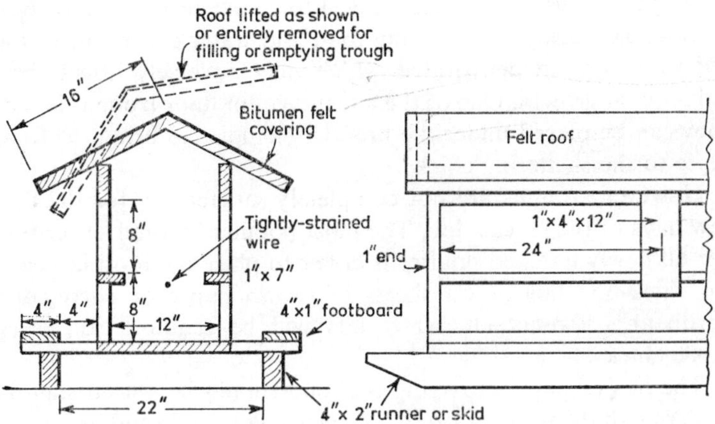

Fig. 10 Turkey range feeder: *left* cross-section, and *right* part side view

Where light is provided in the house by means of windows, it is best to arrange the floor feeders at right angles to the source of light to ensure they are evenly illuminated and to encourage food consumption.

It is important to impress upon the attendant the rule that fresh food should not be put in the troughs on top of old. Unless this rule is adhered to, stale food will accumulate and remain uneaten. When the troughs are replenished, they should be tilted so that the old food slips to one end; any litter is removed and the fresh food is placed in the empty end of the feeder. Frequent filling tends to stimulate food intake.

Large-capacity food hoppers holding from 50 to 100 kg (1 to 2 cwt), perhaps enough for a week's supply, are used by some producers. They are not generally recommended, but are satisfactory if well designed and managed. They are more suitable for use with turkeys than with chickens. Careful adjustment is necessary to prevent waste but there is always a danger of stale food accumulating.

Feeding young chicks

Young chicks should be encouraged to commence eating as soon as possible. Delay in providing food and water can result in the chicks becoming dehydrated which may adversely affect their subsequent growth. One of the advantages of floor troughs is that they can be placed under the brooder canopies to bring the food close to the sheltering chicks.

However, troughs are not completely satisfactory for the first few days of the chicks' life. The most popular method of feeding for all newly hatched poultry is either to place the food in heaps on chick-box lids or on sheets of brown paper or corrugated cardboard. Twenty chick-box lids will be required for every 1,000 chicks.

The lids or papers carrying the food should be placed slightly underneath the brooder canopies. Each day they should be moved further away and towards the end of the first week the normal feeders can be introduced. Some tubular feeders are manufactured so that their bases can be separated and used as flat troughs for the first week or so of the flock's life.

Tubular feeders

Tubular feeders are normally suspended from the roof of the house by means of adjustable guy ropes. They are first used at floor level but are then gradually raised, keeping them at the right height to prevent wastage and to keep the litter from soiling the food. The correct height is just about the level of the backs of the chickens.

There is a tendency to allow too much food to collect in the trough of the feeders. Adjustment of the flow from the hopper is by means of a screw and the tube should be kept screwed down so that there is never more than 0·5 cm of food in the trough.

Unfortunately many designs of tubular feeders are unsuitable for mash feeding, and only those of correct design should be purchased.

MECHANICAL FEEDERS

The choice of food receptacles is between tubular feeders suspended from the roof of the house, hoppers on the floor, bulk hoppers and fully automatic methods of food supply. In large-scale units automatic feeders should be given careful consideration in spite of their greater initial cost.

Food may be delivered mechanically to all types of poultry feeder and there is a choice of a number of different systems. Automatic systems are very labour saving and may be more efficient than hand-feeding methods. A single automatic feeding unit will usually service up to 20,000 birds in one house.

Automatic systems must enable the food, whether in the form of mash or pellets, to be evenly distributed to all feeders, without waste or separation and with the minimum of dust. The system must be designed for easy dismantling, or alternatively arranged so that it can be raised to roof height, to allow easy cleaning of the house. In addition it should not be possible for the birds to perch on the feeders. Ideally the system should allow weighed amounts of food to be supplied so that, if required, a restricted feeding system can be followed. To fulfil these functions the unit must receive regular checking and maintenance.

Chain feeders

Chain feeders were the first of the automatic systems to be introduced and were designed in relation to floor troughs. If the troughs are placed end to end, then the chain conveyor runs the full length of the troughs carrying food from a storage hopper and returning to gather further supplies. The movement of the food is said to attract the birds and encourage consumption. One of the disadvantages of the system is that the food may build up and become wasted at the angles where the run of the feeder changes direction. Another problem is the difficulty of preventing floor-housed birds from standing on, and in, the troughs. The system is well adapted to bringing food to the long thin troughs which run in front of battery cages.

The main supply bin of the feeder usually stands in the food store of the building and this in turn may be filled from a bulk bin by means of an auger, or manually by sacks.

Conveyor tube with flexible wire and discs

Food is supplied through a tube about 57 mm ($2\frac{1}{4}$ in) in diameter containing a steel wire to which are fixed discs which pull the food along. This type of feeder is very popular, and will move up to 500 kg of pellets or mash an hour.

Openings in the tube allow the food to be released either into a continuous trough below it or through downspouts to a series of tube feeders or pans (see Plate 12). The system is equally suitable, therefore, for broilers on the floor or for layers in cages.

One advantage of this system is that the tubes can be arranged to follow the contours of the building, and may be supplied directly from a bulk bin situated outside the house.

Conveyor belt with augers

The food is carried along a tube of about 70 mm ($2\frac{3}{4}$ in) diameter by means of an auger; holes are arranged in the tubes above the specially designed feeding pans for the downward passage of the food. The best feeders of this type have flexible augers which can be up to 50 metres in length.

The system is very satisfactory for breeding birds housed on litter and also for turkeys and ducks.

Food may be delivered at a rate of 800 kg per hour. A flexible auger can be arranged to convey the food from bulk silos outside the poultry building to the supply bins inside.

Movable hoppers

The systems above carry the food to the receptables and are always in some danger of both over-supplying and under-supplying some parts of the system. One alternative possibility is to mount the tube feeders, not from the roof of the poultry house, but from a monorail mounted above normal working level. The feeders can then be carried round the track each to be filled in the food store (perhaps with a carefully prescribed amount of food) before being returned into the poultry house. The system gives less wastage but presents some problems in houses that are subdivided into pens.

Automatic feeders for battery cages

Some of the systems already described can be applied to the feeding of birds in battery cages. When so used they retain the concept of the continuous trough (whether this be the troughs on separate tiers or for all tiers taken over the unit as a whole). For cages, though, the idea of a moving hopper is more common.

Perhaps twenty years ago the so-called 'cafeteria' battery (an interesting application of the restricted feeding opportunity principle) received a good deal of attention. More recently moving hoppers (either fixed to the cage unit or moving independently in the intervening gangway) have been used to supply a continuous trough at regular intervals throughout the day under the control of a time switch.

The most recent development has been to return to the idea typified by the individual cage trough method of the early batteries of separate cage feeding. With the metering devices now available a normal long trough can be divided up into 'cage' lengths and amounts of food (appropriate to the number of birds) can be dispensed outside each cage. This system has already been described on page 208.

Controlling food wastage

Considerable waste can occur from food hoppers, perhaps 20 to 30 per cent of that on offer, and the alert poultryman should be on the lookout for waste at all times. Mechanical feeding systems require careful adjustment to prevent losses.

If unchecked, wastage of food can turn what would otherwise be a successful poultry venture into an unprofitable one. Perhaps the most important single item, where tubular feeders are used, is to keep the feeders really well screwed down at all times. It is necessary to impress on the attendant that 5 mm of food in the troughs is ample. He should resist the tendency to tap the feeders in order to shake down more food into the trough; this action is quite unnecessary and can soon lead to considerable waste.

Waste often results from overfilling, and for this reason tubular feeders should never be more than half-, and flour feeders more than one-third filled. Any sign of wastage should be investigated immediately.

A frequent cause of waste is the failure to raise the height of the feeders as the birds grow. The height of the lip of the feeders should always be level with the backs of the birds.

Improperly adjusted feeders can be another trouble spot and it is wise to make regular checks, particularly at the corners of automatic units.

Infestation by rats and mice can be another source of wastage, and a regular policy for extermination of these pests is well worth while.

Water supply

It is essential for chickens and turkeys to have a reliable supply of clean water at all times, as indicated in Chapter One. The importance of water is clear from the fact that the bird's body is made up of 60 to 70 per cent water.

It has been estimated that the chicken requires 0·91 to 1·36 kg (2 to 3 lb) of water for every 0·45 kg (1 lb) of food consumed. Daily water intake, therefore, varies considerably according to the house temperature and the level of food consumption.

The quality of the water should be high and, if there is any doubt, a sample should be sent for analysis. The water troughs should be cleaned daily, using a suitable disinfectant.

The water supply should always be within easy reach. Thus, during the first week, chicks should not have to walk more than 40 to 50 cm to drink; in the growing and adult stage, never more then three metres.

The water troughs should be placed in position round the brooders the day before the chicks arrive. In this way the water is no longer cold when the chicks commence to drink. In the first few days the troughs may be placed on the litter in order to encourage the chicks to drink. A good deal of water may be spilt and litter may block up the drinkers. To ease this problem, troughs can be placed on cardboard or paper. By the end of the first week, the containers should be raised off the floor.

The height of the drinker from the floor should be adjusted as the birds grow, in order to keep wastage to a minimum. The lip of the trough should be kept level with the backs of the birds.

Round automatic water troughs, designed for use on the floor for 1 to 7 day olds, and suspended above the litter from the end of the first week, are one of the most satisfactory methods of water supply. These troughs can be used for any age of chickens, but turkeys require a more robust design after 6 weeks of age (see Plate 13).

By contrast, long troughs always present problems of level and this particularly when they are fixed on battery cages. It is now more usual to fit nipple drinkers in cage units and simple very reliable valves are now available for these.

Water troughs are available for use with high and low-pressure water systems; the low-pressure drinkers are to be preferred, and

Table 38 Hopper space allowances

Chickens	
First five days	One clean tray per 100 chicks
Growers	2–3 round feeders, or pans, per 100 or 4 cm (1·5 in) of trough space per bird
Adults	4–6 round feeders, or pans, per 100 or 10 cm (4 in) trough space per bird
Layers in cages	10 cm (4 in) trough space per bird
Turkeys	
First five days	One clean tray per 100
Growers	Six round turkey feeders, or pans, per 100 or 8 cm (3 in) trough space per bird
Adults	6 round turkey feeders, or pans, per 100 or 10 cm (4 in) trough space per bird

will require a header tank. The supply system should be fitted with a suitable filter. A medicament proportioner may be fitted next to the header tank to allow medicines and soluble vitamin supplements to be dispensed accurately through the water supply.

Table 39 Average consumption of water per 1,000 chickens daily

Age in weeks	*Litres*	*Gallons*
0–4	20–90	5–20
5–8	90–160	20–35
9–12	160–180	35–40
13–16	180–220	40–50
17–20	220–250	50–55
Adults	220–315	50–70

Table 40 Water-trough space allowances

Chicks and broilers

10 round troughs, 37 cm (15 in) diameter per 1,000
or
10 4–5 litre (1 gall) founts per 1,000
or
2 cm (0·75 in) of trough space per bird

Growers and adults

6 round troughs, 37 cm (15 in) diameter per 1,000
or
2 cm (0·75 in) of trough space per bird

Laying cages

400 nipples per 1,000 birds

Turkey poults

10 round troughs per 1,000 birds
or
2 cm (0·75 in) of trough space per poult

Turkey growers and adults

8 round troughs per 1,000 birds
or
4 cm (1·5 in) of trough space per turkey

CHAPTER TEN

Quality control of poultry foods

Quality in the basic ingredients

INTRODUCTION

The recommended nutrient allowances are not an end in themselves. When used in compiling diets the recommendations must be used in conjunction with a knowledge of the individual feedstuffs available. The final mixture must not only contain all the nutrients in the right proportions, but also be composed of those substances that the fowl will readily eat. The resultant ration must be palatable at all times and, to achieve this, the individual ingredients must be selected and prepared with care.

The quality of the ingredients may be influenced by a number of factors including:

(a) The presence of non-edible items, e.g. silica or harmful weed seeds;
(b) Variation in nutrient content;
(c) The degree of grinding and freshness;
(d) The conditions of storage which can lead to infestations, over-heating or rancidity;
(e) The presence of toxins and pathogenic organisms.

CLEANING INGREDIENTS

Ingredients, particularly the cereal grains, may be contaminated with weed seeds, screenings, straw and silica. In addition foods may contain pieces of metal, like nails or screws, and also string and rubble. Most of these non-edible items will be harmful both to the livestock to which they may be fed and to the mill plant. The ingredients therefore should be cleaned before use. A number of excellent cleaners, also called separators, are available and should be installed in all mills.

Cereal grains should be purchased on sample tests, and when there is a heavy contamination by weed seeds or other matter, the batch should be rejected until it has been cleaned. The weed seeds causing most concern are those of *Crotalaria*, sometimes found contaminating maize and milo and responsible for heavy mortality in young chickens; those of the common vetch; and to a lesser extent, the hairy vetch, which are known to be toxic to turkey poults and may be found in wheat and barley samples.

VARIATION IN NUTRIENT CONTENT

Ingredients vary considerably in feeding value and, wherever possible, they should be purchased on an analysis of their nutritive content. A regular policy should be followed to provide both chemical analysis to establish their theoretical nutritive value, and biological tests using the species to which the ingredients are to be fed, to assess their practical feeding value under carefully controlled conditions, should this prove to be different from what the chemical analysis would suggest.

Cereals, for example, may range from 8 to 14 per cent in their crude protein content, according to the strain used and the fertilizers applied during growing. In the same way, the moisture content of cereals may vary, as a result of conditions at the time of harvesting, from 8 to 15 per cent or even higher.

The protein-rich ingredients not only vary considerably in their calculated protein content, but also in their oil and mineral content. The salt content of fish meal, for example, requires checking and samples that have too much should be rejected.

The nutritive value of the extracted oil-seed proteins, e.g. soya and cottonseed, alters considerably according to the variety, to the manner in which the fruit or seed is processed before extraction of the oil and to the method of oil extraction.

The outer coating of many of the oil seeds is extremely fibrous and if this is included in the meal, the product will be of only poor nutritive value, perhaps unsuitable as a ration ingredient. The process of removing the outer coating is known as decortication. Only decorticated oil-seed residues should be considered for poultry feeding.

Oil is extracted from the seeds by three methods: the hydraulic press, the expeller, and the solvent methods. The last method is the most efficient and involves dissolving out the oil in a standard solvent. The resultant product is then dried and ground and has a residual oil content of 0·5 to 2 per cent.

GRINDING AND PARTICLE SIZE

The fineness to which the food ingredients are milled or ground, i.e. the particle size, is known as the grist of the product, and this has a considerable influence on palatability. Fine grinding gives a particle size of up to 2 mm and coarse grinding particles from 2 to 5 mm.

An unsatisfactory grist may render a ration unpalatable even when it is composed of perfectly good feedstuffs. Raw materials vary in the manner in which they should be ground and the grist of some ingredients is of great importance.

Wheat may be used to illustrate this point. When whole or coarsely ground wheat is used, the diet is very palatable, but when finely ground wheat is included its floury nature may render the mash inedible because of the starchy paste it makes when mixed with the bird's saliva. Wheat, therefore, should always be coarsely ground when it is to be included in the diet.

Barley and oats are usually best if finely ground, using a screen not larger than 1·6 mm ($\frac{1}{16}$ in). The reason for this is that otherwise the fibrous hulls may cause digestive troubles in the stock. The palatability of barley can be greatly improved by the addition of tallow.

Limestone also may cause problems of unpalatability if it is too finely ground when included in a ration. It is for this reason that limestone granules are often preferred for use in diets for laying birds.

To ensure the most satisfactory results, it is generally best to grind the individual cereals separately rather than grinding them all together. If the final mixture appears to contain rather too much dust due to the unavoidable use of fine or dusty ingredients, the inclusion of molasses or fat may help to give a better texture.

MICRONIZATION

Micronization is a form of processing cereals and leguminous seed that involves the infra-red radiation of the grains. In the process the grains are subjected to very high temperatures for about forty-five seconds, and this causes a change in the starch granules. The effect of this is to act as a sort of pre-digestion process and to improve the energetic value of the material by as much as 8 to 10 per cent.

Much work is being conducted to establish the value of micronization of ingredients for poultry feeding but no generalized recommendations are yet available.

STORAGE OF FOOD

The quality of ingredients, of some additives and of whole rations can be greatly influenced by the manner in which they are stored. Improper storage can lead to the development of infestations, over-heating and the growth of moulds, to contamination by rodents and to rancidity.

It is advisable to store all feedstuffs in vermin-proof bins, for if rats and mice gain access large quantities of food will rapidly be rendered at best unpalatable and, at worst seriously infected with disease organisms, particularly those of the salmonella group.

The moisture content of ingredients at the time of storage is of considerable importance. Cereals harvested during wet weather may have a high moisture content, e.g. up to 22 per cent instead of the more usual 8 to 15 per cent. Over-heating and moulds may occur if cereals with a moisture content of over 15 per cent are put into store. Cereals high in moisture should be dried before storage; a number of efficient methods of drying are available.

The use of propionic acid salts has been shown to be of value in controlling the development of moulds in high-moisture grains.

Arrangements should always be made to ensure a rapid 'turn-round' of feed ingredients stored in silos and this can be arranged entirely automatically. In this way ingredients that are somewhat moist can often be dried so that over-heating is prevented. The conveyors should be arranged to allow the food to be drawn from one silo, and after passage round the system to be returned to either the same silo or to any other which is empty at the time. Any 'caking' which may have started in the food will be broken up in this way.

RANCIDITY

Rancidity of food is frequently a problem when ready-milled ground cereals and cereal by-products are purchased as a prelude to making-up a ration. The best safeguard is clearly to purchase only whole cereals and to grind them as required.

Rancid maize meal is a particular cause of trouble under commercial conditions and is often associated with the product having been shipped through tropical areas. Rice polishings are another food ingredient that is subject to this problem. Fats also may turn rancid and vitamin mixtures are liable to loss of value if rancidity of other ingredients occurs.

Rancidity in a meal develops relatively slowly at first, but thereafter the deterioration may be very rapid. The addition of an antioxidant to fats and vitamin mixes retards the development of the first stage but not of the second. Antioxidants, therefore, should be added to the product as soon after preparation as possible.

It is the fats and oils present in the ground materials that are primarily responsible for the development of rancidity. Fats are compounds of glycerin and fatty acids and, in the right conditions of moisture and in the presence of ferments, they will commence to break down into their two constituent parts. This type of rancidity is known as hydrolytic rancidity and is characterized by the presence of free fatty acids. The action of oxygen from the air on the free fatty acids may then give rise to other compounds, the products of oxidative rancidity.

The level of free fatty acids is used as a measure of the degree of rancidity of a food, but this is not a guide of great value unless the original level of free fatty acids of the food is known. Poultry can utilize free fatty acids as efficiently as fats.

Rancidity can soon lead to the loss of, and finally to the complete destruction of the fat-soluble vitamins contained in a food (vitamins A, D and E) and also to the loss of its carotenoid and xanthophyll pigments. Rancid foods may also lead to digestive upsets.

TESTS FOR RANCIDITY

The chemist has two main tests that indicate the quality of a feedstuff in respect of rancidity. They are by no means infallible but are a helpful guide. The first, the so-called lipolytic rancidity test, is based on the percentage of free fatty acids in the oil. The second is based on the peroxides that are formed during the development of rancidity and is known as the peroxide value (PV) measured in milli-equivalents of peroxide per kg of sample.

The FFA of ingredients of animal origin should, for preference, be under 15 per cent and never more than 20 per cent. Thus tallow should have an FFA value under 10 per cent. The FFA of fats of vegetable origin may show considerable variation. An unusually high FFA figure may indicate poor storage or the presence of mites or moulds.

Freshly prepared complete mixes should have an FFA of approximately 20 per cent and should not be fed to youngstock. Foods with FFA of over 35 per cent should not be fed to poultry at all.

Cereal by-products and ground cereals should have a minimum acid value of the oil of 30. Meat meals and fish meals should be purchased on their protein content, their freedom from harmful disease organisms and from rancidity. The PV figure should be as low as possible, e.g. 0 to 10 milli-equivalents. Fats used as food ingredients should have a PV of under 10.

Moulds, toxins and poisons

AFLATOXIN

Stale and musty food is a frequent cause of unpalatability. On no occasion should musty ingredients be ground and included in the dry mash. If stale or musty foods have to be used because there is nothing else, they are best scalded and then used as a small percentage only of a wet mash.

Food ingredients are natural carriers of the spores of fungi or moulds which under favourable circumstances may grow and develop toxic substances known as mycotoxins. These toxins can cause illness in poultry, and sometimes in man, called mycotoxosis. The moulds develop best in cereals and ingredients rich in vegetable protein that have been improperly stored; that is to say, when they have a moisture content of over 15 per cent or more and when the ambient air has a relative humidity of 75 per cent or more.

Aflatoxin is a well-known example. It is produced by the fungus *Aspergillus flavus* and has often been responsible for poor growth and mortality amongst poultry; ducklings and turkeys have been shown to be particularly susceptible. Aflatoxin is often present in groundnut meal and palm-kernel meal and these ingredients should not be used unless they have been shown to be free of the toxin, for which purpose a useful and rapid test has been developed.

OTHER TOXINS

Prussic acid

Improperly prepared cassava meal may contain an unacceptable level of prussic acid; the maximum content should not exceed 1·9 ppm. A precursor of prussic acid is also found in linseed meal but it can be destroyed by boiling the meal before use.

Gossypol

Cottonseed meal may contain a toxin known as gossypol. The maximum level of free gossypol should be 0·01 per cent. Iron sulphate may be used to prevent the gossypol from giving discoloured egg yolks.

Toxins in crotalaria and other legumes

Crotalaria is a legume grown for the purpose of land improvement and the seeds are sometimes found contaminating a succeeding crop of maize or milo. The seeds may generally be removed from the cereal grains by cleaning with the appropriate size screen. They are heart-shaped and black in colour and are toxic to poultry. If by mischance they are in the ration, at a level of 5 per cent they will cause very heavy mortality in young chickens. The symptoms of uneven growth, ruffled feathers and death may occur two to three weeks after they have been eaten.

Castor beans contain a poison known as ricin, whilst lupin-seed meal contains a toxic alkaloid which can be rendered harmless by boiling or soaking.

The seeds of the common vetch and, to a lesser extent, the seeds of the hairy vetch have been shown to be toxic to turkey poults, and the toxic agent is thought to be divicine.

Symptoms of vetch-seed poisoning include difficulty with breathing, poor growth, lack of muscular co-ordination and finally convulsions prior to death.

Digitalin

A case of poisoning due to the consumption of the common foxglove (*Digitalis purpurea*) was recorded in 1951. Some American Bronze and White turkeys out of a batch of seven-month-old birds showed loss of appetite, stasis of the crop and dilation of the eye pupils. The turkeys ceased to drink and became drowsy. Their crops were still distended some forty-eight hours after the birds ceased to have access to the foxgloves.

Opium

Poppy-seed meal may contain opium.

Goitrogens

Rapeseed meal may contain a toxic principle that has been associated with perosis and reduced egg production. The goitrogens' properties vary very much with the variety grown; some types are much safer than others.

Salt poisoning

The inclusion of too high a percentage of added salt in the diet can lead to severe mortality in young turkey poults during their first few weeks of life. The symptoms include so-called dropsy or water belly in which an excess of water accumulates in the abdominal cavity.

Mercury poisoning

Consignments of cereals that have been prepared for use as seeds may sometimes be offered for use as animal food. Samples that have been treated with mercurial products for the control of fungoid disease should be rejected as these may be poisonous.

Quality-control procedures

THE QUALITY-CONTROL PROGRAMME

Every organization producing poultry foods should develop its own quality-control programme and positive steps should be taken to ensure that the programme is well known to all personnel involved. One person should be given the responsibility to see that the detail of the quality-control programme is strictly maintained. The programme should include:

The establishment of nutritional standards for all ingredients to be used.

A procedure for the clear identification of ingredients in the mill.

The labelling of complete foods.

A sampling and analysis procedure.

Quality-control check list.

INGREDIENT STANDARDS

Standards for inspecting food ingredients are essential in order to provide suitable specifications to control the purchase of the ingredients, so that low-quality or potentially harmful substances may be avoided.

The specifications should show the desired nutrient content of the individual ingredients, of pre-mixes and concentrates purchased and, also, the acceptable limits of variation. In the case of cereals, standards for the physical condition are required. The specifications should cover the points summarized in Tables 41 and 42.

IDENTIFICATION OF INGREDIENTS

The utmost care is necessary in the mill to ensure that all ingredients are separately stacked and identified. The mistaken use of ingredients may lead to unfortunate results. The materials in the bag warehouse, as well as in the micro-ingredient store, must be labelled clearly.

Potentially toxic chemicals, including insecticides and pesticides, should be labelled and stored in an area well away from the food-intake area and the micro-ingredient store, to minimize the chance of their accidental inclusion in a food mixture.

Table 41 Sampling procedures for food ingredients.

Raw materials	*Examinations required*
Whole cereals	Inspection, moisture, protein, fibre occasionally, weed seeds
Cereal by-products and ground cereals	Inspection, moisture, protein, rancidity
Animal protein sources other than milk products	Inspection, protein, oil, salt, true protein occasionally, moisture, rancidity occasionally Acid pepsin digestible protein test occasionally
Plant protein sources:	Inspection, protein, oil, moisture, fibre occasionally
Sesame	Sand and silica
Soya	Heat test
Cottonseed	Free gossypol
Groundnut	Aflatoxin
Bulk minerals:	
Limestone	Calcium, silica, particle size, magnesium
Phosphate sources	Calcium, phosphorus, fluorine, particle size
Protein concentrates	Inspection, oil, fibre, moisture, ash, amino acids occasionally
Balanced foods	
Meals	Protein, oil, grist routinely, moisture, salt, calcium, phosphorus, ash, additives occasionally or to meet a specific requirement
Pellets	As above but in addition hardness and size

Table 42 Specifications for purchasing samples of feedstuffs

Grains

Grain should be free from weevils and other insects likely to be injurious to stored grain; it should also be free from the fungus ergot. The sample should have the following characteristics:

Odour:	Sweet
Sound grains:	95% minimum
Broken grain and foreign material:	5% maximum
Mouldy grain:	5% maximum
Moisture:	10%–14%

The density of the bulked sample gives a good indication of its value and the following are minimum standards:

Cereal	*g/100 cm³*	*lb/cu ft*
Barley	67	40
Maize	77	48
Oats	47	30
Sorghum	67	40
Wheat	73	46

Proteins

Protein foods should be purchased against a declared protein content. The contract should contain a penalty clause based on a deduction of the unit price of the protein for each 1 per cent in which the product falls below the declared protein content. The unit price is the cost per tonne divided by the declared protein content, thus soya-bean meal purchased at £121 per tonne with a declared protein content of 44 per cent has a unit price of £2.75.

Mixed foods

Freshly mixed foods may be tested for lipolytic rancidity which should show a maximum acid value of the oil of 30.

The total phosphorus figures may be calculated on the basis that in ingredients of plant origin the available phosphorus is 40 per cent of the total.

The moisture level will vary according to local conditions.

Table 42 (*continued*)

The limits of variation allowed by the Fertilizer and Feeding Stuffs Regulations in the United Kingdom in respect of mixing foods are:

Oil	0·75% or 1/10th of the amount stated which ever is the greater
Protein	1/10th of the amount stated
Fibre	Plus or minus 50% of the amount stated

LABELLING COMPOUNDED RATIONS

Labelling of mixes produced is required to prevent mistakes both in delivery and in use on the farm. The veterinarian responsible for the health of the livestock requires information on the nutrient content of the food and on the drug inclusion to assist him in his task of disease prevention and control.

The labels may be stitched into the bags at the time of sealing without interfering with the rate of bagging. They should include the following information:

- The name and purpose of the food;
- The net or gross weight;
- The calculated nutrient content;
- Details of any special additives specifying:
  - Chemical name;
  - Common name;
  - Level of inclusion in grams per tonne or grams or milligrams per kilo;
  - The purpose of the drug;
  - Any special instruction regarding the use of the product.
- The common names of the food ingredients;
- Any special instructions regarding use including:
  - Age of stock for which the food is designed;
  - Any special precautions involved if the misuse of the product could be harmful to man or livestock.
- Warning, if necessary, regarding any withdrawal period;
- Batch number;
- The name of the manufacturer.

SAMPLING AND METHODS OF ANALYSIS

Constant monitoring of the quality of the food ingredients and completed foods is necessary. To achieve this a regular programme of sampling and analysing the foods is required.

Samples for analysis should be representative of the batch of food, and this can only be achieved by following a carefully standardized routine procedure. The frequency of sampling varies according to need. Details of a suitable sampling procedure are given in Tables 43 and 44.

Physical analysis

All ingredients and foods purchased should be subjected to physical analysis when they arrive. This involves inspection for general appearance and odour and for signs of rodent contamination. In the case of cereals the points to note are the size and condition of the kernels and the presence of moulds and insects, weeds, seeds and foreign materials.

Chemical analysis

Laboratory services are required for conducting the following chemical determinations:

The routine analysis of fats, fibre, minerals, crude protein and testing for rancidity;

The determination of vitamin and amino-acid levels and those of other additives;

Testing for the presence of toxins and pathogenic organisms.

Table 43 Frequency of sampling of ingredients and completed foods

At the start of a mill's operation, the following frequency of sampling is recommended. Experience will allow a selective sampling procedure to be established and the ingredients liable to variation will soon be identified.

Table 43 (*continued*)

All deliveries

All deliveries to be examined to ensure that the goods are free from significant damage and infestation.

Cereals and ground cereals

(a) Large consignments	Sample each delivery
(b) Small consignments	Sample one in ten

Animal- and vegetable-protein sources

(a) General	Sample and test one in ten consignments
(b) Specifically:	
Meat meals, meat and bone meals	Test and sample each consignment
Cottonseed meal and cake	Inspect and sample each consignment and test for free gossypol Analyse one in ten consignments for the other tests or as experience dictates
Soya-bean meal	Sample and test each locally produced consignment for completion of heat treatment, moisture and protein. Reduce rate of testing as experience dictates

Minerals

Bulk minerals	Sample one in ten
Trace elements and additives	Sample each consignment

Balanced foods

Generally about 2 per cent of the production should be sampled. Greater attention may be required for special foods for young-stock and intensively kept animals.

Table 44 Number and size of samples

Samples should generally weigh 1·0 kg and should be stored in moisture-resistant containers, e.g. polythene-lined bags, glass

Table 44 (*continued*)

bottles or airtight tins. The samples should be clearly labelled and dated.

If the batch of food does not appear to be uniform in nature it should be divided into lots and samples taken from each lot.

Number of samples: Food in bags

Number of sacks	*Sacks to be sampled*
Up to 20	4
20–60	6
60–100	8
100–400	10
over 400	20

Collect the sample from at least 15 cm (6 in) below the surface by plunging the closed fist into the food, opening to obtain the sample, reclosing and drawing out. Mix samples together and pile into a cone, flatten, divide into four equal parts and reject three. Repeat the process until the final 1 kg sample is obtained.

Feed in bulk

Collect samples from different areas of the batch, cone, quarter and repeat until the final sample of 1 kg is obtained.

Weight of bulk	*Number of portions*
1–2 tonnes	6
2–5 tonnes	12
5–25 tonnes	20
over 25 tonnes	40

Biological testing

The value of the information provided by chemical analysis is limited because the reports are calculations based on chemical tests rather than from feeding trials with the stock concerned.

On the other hand, feeding trials are time-consuming and, in view of the volume of food used, it is impractical to check the quality in this way before purchase.

When they are used, feeding trials must be conducted with

sufficient numbers and groups of birds according to well-defined statistical procedures. Only in this way can any differences that may appear be said to be significant rather than due to chance. Comparing the birds in one half of a house, for example, with the birds in the other half (the so-called half house test), can lead to the most misleading results.

Quality-control check list

A check list may be of considerable help to management in maintaining quality and also when investigating complaints. An example of a check list and complaints procedure is detailed below:

Mill Check the arrangements to prevent the confusion of one ingredient with another.

What happens to the feed from broken bags?

Check for possible contamination of food by birds and rodents.

Check for infestation in cereals.

How are the sweepings and dust disposed of and, in particular, the waste from the micro-ingredient store?

Are the separators working correctly?

Check for general cleanliness.

Delivery Check the precautions against mixing of foods delivered in bulk.

Check that the bagged food is properly labelled.

Sampling Check the register of samples sent for analysis.

Check sampling procedures.

Formulations Check the formulations register and procedures.

Check the operation of the punch-card system if one is installed.

Micro-ingredients Check that the inventory is up to date.

Are the micro-ingredients properly labelled?

Are the ingredients stored to prevent mixing?

Check temperature record.

APPENDIX ONE

Alphabetical list of poultry food ingredients

Acorns (*Quercus robur*)

The fruit of the oak tree, acorns, may be gathered, dried and ground for use by the domestic poultry keeper at the rate of up up to 30 g per adult bird daily.

Acorn meal is a carbohydrate-rich food but is not suitable for inclusion in specialist mixtures.

The consumption of freshly germinated acorns by birds on range has been associated with brown discoloration of yolks, particularly in duck eggs.

Alfalfa (*Medicago sativa*)

See Green foods.

Algae

Research work concerning the production of animal protein from Spirulina is encouraging. Spirulina is one of the muticellular algae or water plants and work is being conducted in Algeria, Egypt and Mexico on collaboration with the Institut Français du Petrole (I.F.P.).

NUTRITIONAL VALUE

Spirulina has formed part of the diet of certain tribes of Chad for many centuries and is said to have been used by the Aztecs near Lake Texooco near Mexico. The protein content is around 70 per cent on a dry-matter basis with a good balance of amino acids. The product is rich in B_{12}, carotenoids, particularly the xanthophylls, and has 4·1 per cent nucleic acid. Feeding trials with animals are said to have been satisfactory.

PRODUCTION

Spirulina may be grown in open basins and the rate of growth depends upon the light flux and temperature. Optimum temperature required is 30° to 35° C, when production is 16 g dry matter per m_2/day. Production in Algeria averages 8 g/m_2 daily due to the Spirulina being dormant during the winter.

Bacterial protein

See Single-cell protein

Barley (*Hordeum vulgare*)

Rate of inclusion in poultry diets:

For youngstock	5 to 20 per cent
For adults	20 to 50 per cent

Barley, one of the most widely grown cereals, is a carbohydrate-rich food. It should be finely ground before use, since poultry do not find whole barley very palatable probably because of the barley hulls and awns. Fine grinding breaks these fibres into pieces too small to give trouble. Poor-quality samples of barley, with badly filled grains, should be avoided because of their high fibre content. Barley meal is not generally included in broiler feeds for this reason but the value of the moderate samples may be improved by pelleting.

The addition of fungal and bacterial enzymes to certain types of barley has been shown to be capable of improving the nutritional value for poultry feeding. The improvement is due to an increased energy utilization of as much as 20 per cent. The response seems to be determined by climatic conditions at the time of harvest.

Barley feed or barley bran

Barley feed is made up of the bran and husk of barley that is recovered during the manufacture of pearl or pot barley. Barley feed is generally unsuitable for poultry feeding because of its high fibre content.

Barley feed may be recovered as fine, medium and coarse grades, and if the fine dust has a low fibre content, e.g. 4 to 5 per cent, its use may be considered for non-specialized low-density mixes.

Beans

Broad, Windsor or Horse beans (*Vicia faba*); haricot or Navy beans (*Phaseolus vulgaris*).

Rate of inclusion:

Youngstock	2·5 to 10 per cent (up to 15 per cent for broilers with methionine fortification)
Growers and adults	5 to 10 per cent

Good-quality freshly ground whole beans have a content of vegetable protein of about 25 per cent. The protein is not of high quality, being low in methionine.

Other varieties of beans of the *Phaseolus* species may sometimes be available but should be analysed before they are put into general use. Some varieties, e.g. Rangoon, Burma or red beans, contain glucosides that may, under moist and wet conditions, be turned into prussic acid. The wild Java bean (*Phaseolus lunatus*) is particularly dangerous to use. (*See* Castor bean meal.)

Biscuit meal

Rate of inclusion:

Youngstock and adults 10 to 40 per cent

Biscuit meal is prepared by grinding biscuit pieces broken during manufacture and is usually available only in small quantities. The meal is highly digestible, rich in carbohydrates and may be used in the diet of poultry of all ages.

The meal absorbs moisture and swells considerably on wetting; for this reason it is more suitable for use in wet rather than in dry mashes. Meal containing chocolate residues may not be satisfactory.

Biscuit meal used in conjunction with fresh milk was, at one time, widely used for small-scale chick and turkey rearing.

Blood meal

Rate of inclusion:

Youngstock and adults 1 to 2 per cent

Blood meal is not an ingredient of choice for poultry and its use has been associated with reduced rates of growth. Good-quality blood meal is, however, one of the richest sources of animal protein, but protein which is of poor biological value for poultry. It is a good source of the amino acids lysine and methionine but almost totally deficient of isoleucine. The protein content is likely to be in the range 80 to 85 per cent.

The use of blood meal tends to make the mixture unpalatable and its use for youngstock should be avoided.

Bones

See Calcium and phosphorus.

Bread

Good-quality bread is a useful carbohydrate-rich food and is best used as part of a wet mash (up to 25 per cent) for all classes of poultry. Bread may have a protein content of 14 per cent.

Brewery and distillery by-products

DRIED BREWERS' GRAINS

Dried brewers' grains are not a satisfactory food for poultry owing to their high fibre content. The grains are a by-product of the brewing industry and are a mixture of barley skins and oat husks. The latter may form as much as 30 per cent of the product.

A similar product, known as dried distillers' grains, is available as a by-product from the production of whisky.

DRIED DISTILLERS' SOLUBLES

Rate of inclusion:
Youngstock and adults   1 to 2 per cent

Good-quality dried distillers' solubles (about 27 per cent protein) are a reasonable source of protein, especially of the amino acid lysine and a good source of some of the vitamins of the B complex.

Dried distillers' solubles are a by-product of the production of malt. The product is somewhat variable and should be purchased only against its known nutrient value.

DRIED DISTILLERY CONCENTRATE

Rate of inclusion:
Youngstock and adults   1 to 2 per cent

Dried distillery concentrate is available either as a by-product from the production of malt from barley, or from the later distillation of whisky. In the former case the product is, in fact, malt and has a protein content of about 38 per cent. Dried whisky distillery concentrate has a protein content of about 44 per cent. Both are good sources of vitamins of the **B** complex.

Another distillery product is known as whisky dreg and has a protein content of about 40 per cent.

MALT CULMS

Malt culms are not a suitable food for poultry. The culms consist of the rootlets and shoots of the germinating barley that have subsequently been dried. On occasions malt culms may contain appreciable quantities of arsenic.

Buckwheat (*Fagophyrum esculentum*)

Rate of inclusion:

Youngstock and adults 5 to 20 per cent

Ground buckwheat has a somewhat higher fibre content, 15 per cent, compared with the common cereals and may be used only in limited amounts. The whole grains are generally considered rather unpalatable and are not readily eaten by fowls.

Calcium and phosphorus

Table 45 Typical analyses of calcium and phosphorus sources

	Calcium %	*Phosphorus* %
Bone meal	24–30	10–15
Chalk	38	—
Di-calcium phosphate	18	20–24
Dried battery waste	3·8–12·5	1·7–2·8
Fish meal	5–9	3–4
Limestone	33–38	—
Meat and bone meal	7–10	2·5–4·0
Oyster shell	38	—

Supplementary calcium in the diet is usually supplied in the form of calcium carbonate and the most popular sources are limestone and oyster shell. Ground chalk, cockle and other marine shells and poultry egg shells may also be used. Calcium and phosphorus are often supplied together, generally in the form of bone meal and di-calcium phosphate.

Other potential sources of calcium, like calcium sulphate, should not be used since they may have inclusions of poisonous minerals such as vanadium.

Similarly, other forms of phosphorus may be available but should not be used until they are known to be suitable for poultry. Unusual sources of phosphorus require chemical checking for the presence of harmful minerals, followed by biological testing with poultry to check the availability of the phosphorus.

LIMESTONE

Good-quality limestone and oyster shell should contain about 38 per cent calcium and not more than 0·5 per cent magnesium. The quality of limestone from any one source tends to remain constant, and its colour, varying from white and grey to brown and red, tends to typify the area of production.

Limestone is available for use in food mixtures as a powder or in granular form. Limestone grits, designed for free-choice feeding by poultry, are available in chick, pullet and adult sizes.

The comparatively high levels of limestone that may have to be included in layers' meals has led to the use of granular forms of limestone in order to overcome the problem of very dusty food mixes. The specification for No. 6 granules, which are preferred for layers' meals, is:

100 per cent to pass No. 7 British Standard sieve.
89 per cent to pass No. 14 British Standard sieve.
4 per cent to pass No. 25 British Standard sieve.
1 per cent only to pass No. 30 British Standard sieve.
Nil to pass No. 36 British Standard sieve.

Bone flour

Phosphorus and also a substantial quantity of calcium are supplied by bone flour. The level of phosphorus ranges from 12 to 15 per cent and the calcium from 24 to 30 per cent.

Satisfactory sterilization and positive steps to prevent re-contamination are essential. Improperly sterilized bone meals have been associated with outbreaks of anthrax in cattle, and for this reason, many animal food manufacturers decline to use bone meals.

Purchases should be made against a certificate of sterilization and there should also be a guarantee of the calcium and phosphorus content.

DI-CALCIUM PHOSPHATE

Di-calcium phosphate is the phosphorus supplement of choice amongst animal food manufacturers and contains 20 to 24 per cent calcium and about 19 per cent phosphorus. It is produced by chemical treatment of bones and rock phosphate. The product should be free from harmful minerals, e.g. the maximum level of fluorine should be 0·2 per cent.

Suggested specification:

Calcium (Ca)	25 per cent
Phosphorus (P)	19 per cent
Fluorine (Fl)	0·2 per cent maximum

Carobs

See Locust beans.

Carrots (*Daucus sp.*)

Shredded fresh carrots may be used by the domestic poultry keeper at a level of 15 to 30 g (½ to 1 oz) per bird daily. Carrots are a rich source of the pro-vitamin A carotene and are a useful source of energy. Contrary to expectations the pigments in carrots do not have any appreciable effect on yolk colour.

Cassava (*Manioc sp.*)

Rate of inclusion:

Youngstock	2 to 5 per cent
Adults	3 to 30 per cent

Cassava meal, also known as tapioca meal and manioc meal, is a readily digestible carbohydrate food that is made from the roots or rhizomes of the cassava plant. In the natural form the cassava rhizomes contain prussic acid, bitter varieties more so than the sweet, but this is largely removed during the course of processing.

The acceptable level of prussic acid in cassava meal is about 2 ppm but analyses should always be carried out to avoid the use of samples which may have a much higher content. Cassava is not a simple substitute for cereal flours because of its very low protein content and low pigmenting ability. It also tends to be very dusty.

Castor-bean meal (*Ricinus communis*)

Castor-bean meal is a by-product from the production of castor oil by the solent process but is not a suitable ingredient for poultry feeding as it contains a poisonous substance known as ricin. The toxin can be destroyed by heat treatment to produce 'detoxicated castor-bean meal'.

Castor beans may be found as a contaminant in other vegetable protein foods produced from oil seeds if due precautions have not been observed at the oil production plant.

Charcoal

Charcoal has no real place in modern poultry nutrition. Poultry of all species will readily consume granular charcoal but the substance has no nutritional value. It has great powers of absorption and, in the past, has often been recommended as a treatment for scouring.

Chickpeas

See Peas.

Citrus pulp (*Citrus sp.*)

Dried citrus pulp is not generally considered as a suitable ingredient for poultry feeding, though often used for cattle. Some success has been observed from including 10 per cent pelleted citrus pulp in layers' rations.

Clover meal

See Green food.

Coconut or copra meal (*Cocos nucifera*)

Rate of inclusion:

Youngstock	2·5 to 5 per cent
Adults	5 to 30 per cent

Good-quality freshly ground coconut meal may be regarded as a useful source of supplementary protein in many tropical areas. Its protein content is usually in the region of 20 per cent and it also contains 12 to 15 per cent fibre.

Coconut meal is a by-product remaining after the extraction of coconut oil. It should not be the only source of protein in the diet, for the use of high levels of coconut meal has been found to lead to a depression of egg weight.

Coconut meal is variable in composition and is relatively high in fibre. Care must be taken to avoid rancid samples. It absorbs moisture readily and should, therefore, be stored in a dry place.

The term 'copra' is used to describe the dried kernels of the coconut and the meal made from it is sometimes called copra meal.

Coconut oil may be used in the same manner as animal fats and at similar levels.

Cod liver oil

See Fish liver oils.

Corn (*Zea mays*)

See Maize.

Cottonseed meal (decorticated) (*Gossypium sp*)

Rate of inclusion:

Youngstock	5 to 10 per cent
Adults	5 to 20 per cent

Good-quality decorticated cottonseed meal is a useful source of supplementary protein and contains 35 to 40 per cent protein. It should have a free gossypol content of not more than 0·01 per cent. Undecorticated meal is not suitable for poultry feeding because of its high fibre content.

The meal is produced from the seeds of the cotton plant after the husks have been removed. It contains a toxic substance known as gossypol and also certain fatty acids. After storage the eggs laid by birds fed on 5 per cent cottonseed may develop olive spots on the yolks due to the presence of the gossypol, or pink discoloration of the albumen caused by the fatty acids.

Gossypol may be removed by steaming the seeds prior to extraction of the oil; new strains of cottonseed are being developed which have a low content of seed gossypol.

The use of iron sulphate in the diet at the rate of four parts to one part of free gossypol, up to a maximum of 400 ppm, will overcome the problem of yolk discoloration.

Table 46 Suggested specification for decorticated cottonseed meal

Analysis	*Solvent*	*Expeller*
Moisture	9%	9%
Protein	40%	40%
Oil	0·8%	8·0%
Fibre	11%	8·0%
Free gossypol, not more than	0·01%	0·01%

Tolerance level of gossypol in food

	ppm	*% ferrous sulphate*
Broilers	200	0·1%
Layer	50	0·1%

Cow peas

See Peas.

Crab meal

See Fish meal.

Dari

See Sorghum.

Dates (*Phoenix dactylifera*)

Dried date pulp meal may be available as a by-product from the preparation of alcohol and date syrup. The protein content may be in the region of 5 per cent but the fibre content is variable according to the type of by-product. Date pulp meal from the syrup industry may have a fibre content as low as 1 per cent but that from the manufacture of alcohol as high as 30 per cent. It is of little use, therefore, in poultry food mixtures, although the inclusion of whole ground dates at 10 per cent of the mash has given satisfactory growth with youngstock.

Distillery by-products

See Brewery and distillery by-products.

Dried poultry waste

See Poultry by-products.

Emmer grain (*Triticum dicoccum*)

See Wheat.

Fats and oils

Rate of inclusion:

Youngstock and adults	Mash	1 to 3 per cent
	Pellets	1 to 7 per cent

Fats and vegetable oils are a rich source of energy that will supply 2·25 times as much energy as carbohydrates per unit weight and are thus the richest natural source of energy. They are also a supplementary source of essential fatty acids.

The normal inclusion rate is 1 to 3 per cent but higher inclusions may be obtained by using special equipment that sprays fat on to pellets. High levels of fat tend to produce soft pellets.

The use of fats allows mixtures with much higher nutrient densities to be produced. Alternatively they can supplement larger quantities of lower-energy ingredients like barley to give a mixture with a lower overall cost.

Fat also helps to reduce the problem of separation of ingrients that sometimes occurs during the handling and delivery of mash in bulk; it speeds the production of pellets and reduces the wear on pellet dies. 1 to 2 per cent fat in a mash can be of considerable help in reducing dustiness and improving the palatability of diets.

The grades of animal fat used for poultry feeding are known as tallows; they should be of good colour and free from objectionable smells. The fat should include an antioxidant at a level of 0·02 per cent, have a free fatty acid content of not more than 7 per cent and a minimum peroxide value of 10 milli-equivalents per kg.

Many refined vegetable oils are also satisfactory for use in poultry feeds including the oil of coconut, maize, sesame, soya and sunflower. Animal fats usually give a firmer fat in the finished carcase than do vegetable oils.

The decision whether or not to use fats or oils in the feed will generally depend upon their comparative price in terms of calories with other vegetable sources. Some fats apparently supply additional energy to that indicated by their ME value (see pages 42–43). Because of this 'heat increment' fat-rich diets are of some help in maintaining energy intake in birds subjected to high environmental temperatures.

Fats should be stored at low temperatures and used as quickly as possible to minimize rancidity and the absorption of taints.

Special equipment is required to include fats in a diet for they have to be liquefied by heating before being sprayed into the mixer.

Feather meal

See Poultry by-products.

Fish liver oils

Rate of inclusion:

Youngstock and adults 0·5 to 2·0 per cent

Stabilized cod liver oil and the oils of certain other fish are good sources of the vitamins A and D_3 and their value is expressed in terms of International Units (I.U.), and British Standard Units (B.S.I.). Thus the best oils for poultry are stabilized and contain about 100 B.S.I. units of vitamin D_3 and 800 to 1,000 units of vitamin A per gram.

When fish liver oils are exposed to the air there is a danger of rapid loss of vitamin A by oxidation. For this reason, the oil should be sprayed or mixed into the mixture immediately before use. Even with this precaution the vitamin A value of the oil is lost within a few days.

Fish liver oils should be stored in airtight containers and not heated. Unstabilized fish liver oils should be used immediately after preparation because of the danger of rancidity.

The use of some fish oils has been associated with taints in eggs and poultry meat, and for this reason the inclusion of fish oils may be restricted to 1 per cent of the total diet.

In view of the development of dry forms of vitamins, fish liver oils are not now generally used.

Fish meal

Rate of inclusion:

Youngstock and adults 2 to 10 per cent

Good-quality fish meal, containing 55 to 66 per cent protein, is one of the most valuable of the animal-protein feedstuffs and may be used as the only protein supplement in the diet of growing and adult poultry. At a level of 2 per cent it will provide all the vitamin B_{12} required and it is an excellent source of many of the essential amino acids, a rich source of choline and a good source of calcium, phosphorus and some of the important trace minerals.

This high level of mineral content may render it unsuitable for use as the sole source of protein in starter diets.

Fish meal can vary tremendously in composition and must be utilized on the basis of its analysis. Protein content may very from 50 to 70 per cent. The maximum salt content should be 7 per cent, 4 per cent is better, and the oil content as low as possible, around 4 per cent.

Good-quality meals can be prepared from a variety of different types of fish, including anchovies, herrings, menhaden, sardines, whole white fish and from fish waste after whole fish have been prepared for human use. White fish meals are considered the best and can be identified by their light colour.

In the UK the term 'white fish meal' can only be applied to

products that meet the specifications laid down in the Fertilizer and Feedingstuffs Act. Poor-quality fish meals are best avoided and used as fertilizers. The use of unsatisfactory fish meals has been associated with taints in eggs and poultry meat and with gizzard erosion in broilers.

CRAB MEAL

Crab meals may sometimes be available with a protein content of about 30 per cent and approximately 15 to 16 per cent calcium. It is made from the residues of crabs following their preparation for human use. It should be used with care because of its high mineral content.

SHARK OR DOG-FISH MEAL

Shark meal is useful but has been shown to have a lower nutritive value than good-quality white fish meal.

Fish solubles

Rate of inclusion:
  Youngstock and adults 1 to 5 per cent

Condensed fish solubles are a useful source of the essential amino acids, of the vitamins, particularly choline, riboflavin and vitamin B_{12}, and may also be a source of an unidentified growth factor.

They are supplied from the fish-meal factory as a brown treacle-like liquid with a protein content of about 30 per cent, or dried with a protein content of about 60 per cent.

The semi-liquid solubles are best used mixed with mash and then turned into pellets, or used as part of a wet mash.

Flax (*Linum usitatissimum*)

See Linseed.

Gram (*Cicer arietinum*)

Rate of inclusion:
  Youngstock 5 to 10 per cent
  Adults 5 to 20 per cent

Gram is a protein rich feedstuff that is found throughout the Indian sub-continent, where it is frequently used as a major part of the diet of domestic poultry. The feeding value of gram is similar to that of beans.

Grape pulp (dried)

Dried grape pulp is sometimes available with a protein content of about 6 per cent and about 10 per cent fibre but is considered unsuitable for inclusion in poultry foods.

Greaves

See Meat meal.

Green food (fresh)

Berseem (*Trifolium sp.*); grass (*Lolium* and *Festuca sp.*); kale (*Brassica sp.*); lucerne (*Medicago sativa*).
Rate of inclusion:
Youngstock and adults 0·5 to 3 kg per 100 birds daily
Fresh green food is a good source of many vitamins and trace minerals but it is rather bulky to handle owing to its high moisture content.

Green food is rich in pigments and laying birds fed on a liberal diet of green food will produce eggs with richly coloured yolks. It must be used whilst young for the crop rapidly increases in fibre content and declines in nutritional value as it becomes older.

With birds kept extensively, well-managed pastures and leys may provide excellent poultry food during the spring and early summer months for poultry of all ages. Birds on free range over short, good-quality grass and clover may consume 5 to 10 per cent less purchased foods than other stock. However, once the pasture has been allowed to grow past the flowering stage it becomes almost useless to poultry.

Green food is often used under intensive conditions to prevent boredom and subsequent outbreaks of feather-picking and cannibalism.

Fresh lucerne and berseem are first-class green foods for poultry and, if properly managed, will yield four cuts per year for four to six years in succession. The yield may greatly be increased by the liberal use of poultry manure.

Kale is another useful green crop.

For an all-the-year-round supply of green food, lucerne may be grown in season to cover a period of six months, followed by kale for the rest of the year.

Turkeys and ducks will consume large quantities of green fodder but it is not always desirable to allow them unrestricted access. During the early season when the green food is of high quality, consumption need not be controlled, except in the case of ducks, if too much pigment is deposited in the skin. Later in the season the nutritional value of the fodder may be considerably lower and will not sustain adequate growth rates.

Green food (dried)

Alfalfa meal (*Medicago sativa*); berseem (*Trifolium sp.*); dried grass meal; Bokhara sweet clover (*Melilotus alba*).

Rate of inclusion:

Youngstock	2·5 to 5 per cent
Adults	2·5 to 10 per cent

Dried lucerne meal (alfalfa), berseem meal and grass meal are valuable feedstuffs for poultry. The protein content is variable in the range 15 to 20 per cent but the products are a good source of vitamins and some trace minerals. They are a rich source of the pro-vitamin A and riboflavin and also supply vitamin K and folic acid.

An inclusion of at least 2·5 per cent dried grass meal or alfalfa in the diet of breeding and egg-producing hens is a frequent recommendation.

Green fodder meals may also be used as a source of pigment in broiler foods where a deep yellow colour in the skin of the broiler is desired, and in layer diets to provide yellow yolks.

The nutritional value of dried fodders depends to a large extent upon the stage at which the crop is cut; the younger the grass the better. Dried green-food meals should be purchased, therefore, on their nutritional content. A good-quality product should contain not less than 250 mg/kg of active pigmenting xanthophylls, the fibre level should be not more than 21 per cent and the sample should contain at least 20 per cent protein.

A good sample should have a fresh green appearance but colour is not in itself indicative of the nutritive value of the meal.

LEAF PROTEIN

Leaf protein is prepared by first extracting the juice from leaves and then extracting the protein from the leaf juice. A commercial process of preparing leaf protein is being developed and the meal could become a valuable source of supplementary protein. The residues following the extraction of the juice, but with added nitrogen, may well be suitable for feeding to ruminants.

Lucerne leaf protein can be included at the rate of 10 per cent of the ration for layers. Higher inclusions in some cases have caused a sharp fall in egg production, perhaps because of the pressure of saponins in the concentrate.

Groundnut meal (*Arachis hypogaea*)

Rate of inclusion:

Youngstock	5 to 10 per cent
Adults	5 to 20 per cent

Groundnut meal is also known as arachis nut, earthnut, monkey nut and pea-nut meal. It may be used in diets for poultry of all ages and species to provide up to 75 per cent of the protein supplement of the diet. It cannot constitute the whole of the protein additive since it is deficient in some of the essential amino acids notably lysine and methionine.

Poor-quality groundnut meal may contain a toxin known as aflatoxin. Ducks and turkeys are particularly susceptible to its effects and may suffer heavy mortality.

Only decorticated groundnut meal that has been tested and found free from aflatoxin should be used. Expeller groundnut and undecorticated groundnut is quite unsuitable for poultry feeding. It is also wise to avoid groundnut meals that have become overheated during preparation.

Guar (*Cyanopsis tetragonolaba*)

Guar-germ meal, after detoxification, has been offered as a protein supplement. Too little is yet known about its feeding qualities to offer it as a general recommendation.

Hemp (*Cannabis sativa*)

Rate of inclusion:

Youngstock	5 to 10 per cent
Adults	5 to 20 per cent

Hemp meal is a by-product from the manufacturers of rope and has a similar feeding value to that of cottonseed meal and groundnut meal.

Illipe nuts (*Shorea stenoptera*)

The extracted oil seed meal of Borneo illipe nut is not suitable for poultry feeding though it is used for cattle.

Kapok (*Ceiba pentandra*)

Rate of inclusion:

Youngstock	5 to 10 per cent
Adults	5 to 20 per cent

Kapok seed meal is similar in feeding value to cottonseed meal and groundnut meal and is prepared following the removal of the kapok or floss.

Kitchen waste (processed)

Rate of inclusion: *Per bird/day*

Youngstock	30 to 60 g (1 to 2 oz)
Adults	60 to 80 g (2 to 8 oz)

Processed kitchen waste collected from hotels and catering establishments, processed at sterilizing temperatures, can be a satisfactory poultry food when used as part of a wet mash. The waste is best steam cooked, processed and poured into moulds containing about 25 kg. On cooling the waste sets into a semi-solid mass that has the consistency of a stiff pudding.

In this form it has a high moisture content and may be fed as part of a wet mash. It may, however, be dried and ground and then becomes a reasonably rich protein feedstuff. The material should be purchased on analysis, and it may be used to form up to 15 per cent of the diet of rations for growers and adult stock.

Leaf protein

See Green food.

Lentils (*Lens esculenta*)

Rate of inclusion:

Youngstock	5 to 10 per cent
Adults	5 to 20 per cent

Lentil meal is an unbalanced protein-rich food of similar feeding value to beans.

Linseed cake meal (*Linum usitatissimum*)

Rate of inclusion:

Growers and adults up to 2·5 per cent

Extracted linseed cake meal is not a satisfactory ingredient for poultry foods owing to the presence of a cyanogenetic glucoside from which prussic acid may develop if the linseed becomes damp at the correct temperature of fermentation. The glucoside can be destroyed by boiling the linseed for ten minutes.

Linseed cake meal is made from the residues that follow the extraction of the oil from the seeds of the flax plant. It is best used as part of a wet mash and is a good source of the amino acid trytophan.

Liver meal

Rate of inclusion:

Youngstock and adults 1 to 5 per cent

Liver meal is a first-class protein-rich food made up principally from whale livers, animal livers that are unfit for human use and from the residues obtained during the manufacture of liver extract.

It is a good source of many of the essential amino acids including cystine, glycine, leucine, lysine, and methionine. It is also an excellent source of the vitamins of the B complex, particularly choline, riboflavin, niacin and vitamin B_{12}. Liver meal may also contain an unidentified nutrient known as the 'whey factor'.

Care should be exercised when storing liver meal because, owing to its oil content, it is inclined to become rancid.

It is necessary to distinguish between true liver meal and liver meals that contain other abattoir by-products including lungs and hearts. The latter, sometimes called liver and lung meal, may be regarded as a good source of amino acids and B complex vitamins, but less so than the pure product.

Locust beans (*Catonia siliqua*)

Locust beans are also known as Carob beans and are not suitable for poultry.

Lucerne

See Green food.

Lupins (*Lupinus sp.*)

Lupin-seed meal is generally regarded as unsuitable for poultry feeding owing to the presence of a poisonous alkaloid. The toxin can be removed by boiling or soaking.

New varieties of lupins have been developed which are relatively free of alkaloids and these have been used at a level of 15 per cent in the diet of youngstock. The new varieties of lupins have yielded meals that have ranged in protein content from 30 to 40 per cent.

Maize (*Zea mays*)

Rate of inclusion:

Youngstock and adults 10 to 70 per cent

Two distinct types of American corn or maize, white and yellow, are grown. The yellow form is the more popular and differs from white maize only in the level of yellow pigments in the outer skin of the grain's pericarp.

For feeding, maize is normally ground, is low in fibre, and is the richest source of energy available for poultry diets with the exception of fats and oils.

Maize oil is likely to become rancid once the maize is ground and the use of freshly ground maize is the only way to avoid the problem. Imported maize meals should be avoided for the same reason.

About 55 per cent of the maize oil is linoleic acid, thus with an oil content of 4 per cent, maize used at the rate of 45 per cent of the diet of layers and breeders will meet all the requirements of the bird for linoleic acid to help maximum egg size.

Blighted maize should not be used for poultry feeding until its value has been established by a feeding test with a small number of birds. Following a corn blight epidemic in the USA in 1970, which reduced the total crop to only 70 per cent of normal the use of blighted corn for poultry feeding was investigated. Maize in which 25 per cent of the kernels were blighted with the fungus *Helminthosporium maydis* was fed at a level of 55 per cent to broilers with no detrimental effects becoming apparent.

On the other hand the use of experimentally moulded maize has been shown to result in severe leg deformity in young chicks by twelve days of age, together with poor food conversion and reduced growth.

Maize contaminated with *Fusarium roseum*, a fungus that can develop during storage, may lead to a reduction in egg-shell quality.

The protein of maize consists mainly of zein which is of low nutritive value and which is deficient in the amino acids lysine and tryptophan, but recently plant breeders have been successful in developing strains of maize that have increased values for lysine compared with standard strains, i.e. 4·53 to 4·87 per cent lysine compared wuth 2·98 to 3·03 per cent. The gene controlling the lysine content of maize is called opaque —2.

Maize by-products

Rate of inclusion:

	Youngstock	*Adults*
Maize-germ meal	2·5%–10%	5%–25%
Maize-gluten feed	2·5%–5%	5%–25%
Maize-gluten meal	2·5%–5%	5%–25%
Maize grits	10%–70%	10%–70%

The production of maize starch results in a number of by-products that can be used for animal feeding; these include maize-germ meal, maize-gluten meal and maize-gluten feed. The starch that is produced is used for corn flour, glucose, and for the production of household starch.

The first process in the manufacture of starch is to remove the maize germs which are then available for grinding into maize-germ meal. The de-germed grains are then ground and stirred in water where the starch sinks and the maize gluten floats. The difference between maize-gluten feed and meal is that the gluten feed includes the maize bran.

Maize-germ meal is a good source of vitamin E, and contains about 20 per cent protein, has a high oil content and, for this reason, freshly ground meals only should be used in order to avoid the danger of rancid samples.

Maize-germ meal may also be available as a by-product from the brewing and distillation industry; this type of meal, however, usually contains only 10 per cent protein.

Maize-gluten meal is a somewhat better product than the feed, and has a higher protein content – about 40 per cent, and a lower fibre content – about 4·5 per cent.

Maize-gluten meal is a good source of the pro-vitamin A, methionine and the pigment xanthophyll.

CORN COBS

The corn cob that remains after the maize berries have been separated from the seed heads are sometimes ground to become corn-cob meal. Corn-cob meal has no value for poultry feeding.

FLAKED MAIZE

Maize is sometimes cooked and rolled to produce a product known as flaked maize. The process increases the digestibility of maize but the production costs are higher than the increase in feeding value.

HOMINY FEED

Hominy feed is a by-product of the dry milling of maize and

consists of the bran, germ and attached portions of the starchy part of the maize. Hominy feed has the highest average content of linoleic acid (3·3 per cent) of any maize product and is a valuable poultry food.

MAIZE GRITS

Maize grits consist of very fine pieces of maize that arise as a by-product when maize is broken or kibbled.

Maize grits are particularly favoured by the domestic poultry keeper for feeding to young chickens during the first eight weeks of life.

Malt culms

See Brewery and distillery by-products.

Manioc meal

See Cassava.

Meat meals

Meat meal; meat and bone meals.
Rate of inclusion:
Youngstock and adults 2·5 to 10 per cent

Meat meals are generally made by cooking, drying and grinding meat that is unsuitable for human use. Meat and bone meal comes from the whole carcases less hooves and skin of farm animals that have died.

Both products are very variable in quality. In the United Kingdom, meat offal meals with 55 per cent protein or over and a maximum salt content of 4 per cent can be described as meat meals; products with less than 55 per cent protein are classed as meat and bone meals. For poultry feeding the latter should contain not less than 50 per cent crude protein.

Unless these products are properly sterilized there can be a danger to health. They must be subjected to a temperature of 100° C for not less than one hour.

Greaves, a low-grade meat meal, is a by-product from the tallow industry and comprises the membranous matter that is separated from the animal fat on melting. Greaves are usually only available in small quantities, and may be used as 5 per cent of the diet in grower and adult rations.

Microbial protein

See Single-cell protein.

Milk

Rate of inclusion:

Dried separated milk Dried buttermilk Dried whey	Youngstock and adults 2·5 to 5 per cent

Milk is an extremely valuable protein food which may supply most of the bird's protein requirements.

The protein of milk is almost completely digestible, has an extremely high biological value and supplies all the essential amino acids except glycine. Milk is a good source of vitamins of the B complex, particularly riboflavin, pantothenic acid, B_{12}, and some choline. In addition, milk is a useful source of some trace minerals including iron and manganese.

Milk and milk by-products may contain an unidentified nutrient known as the 'whey factor'.

Liquid milk is rather difficult to use under commercial conditions but may be used by domestic poultry keepers. Allowing the bird's milk only to drink is equal to adding 13 per cent of dried milk to the diet.

Liquid milk should be supplied either fresh or sour but consistently one or the other. Digestive troubles may follow if a change is made from one type of milk to another.

Goats' milk may be regarded as having a similar feeding value to cows' milk. Buffalo milk, on the other hand, is very rich in fat, about 6 per cent, and should be watered down to at least 3 per cent fat before use.

It is best to use separated liquid milk rather than whole milk because the fat tends to stick to the mouths and beaks of the birds.

Milk by-products

Rate of inclusion:

	Youngstock and adults:
Buttermilk and separated milk	ad-libitum *supply*
Dried separated milk	2%–5%
Cheese waste meal	5%–20%
Whey paste	2%–10%
Dried whey paste	2%–5%

BUTTERMILK AND SEPARATED MILK

Buttermilk and separated milk are by-products from the manufacture of cheese and are available for poultry feeding either in their liquid or dried state.

Dried separated milk is a valuable source of protein, many of the vitamins of the B complex and some minerals.

CHEESE MEAL

Cheese meal is a valuable high-quality protein in food that may occasionally be available in small quantities. It is extremely palatable and is readily eaten by all classes of poultry. Care should be taken to ensure that the cheese meal is not confused with other by-products from the cheese factory as some of these may have a high salt content which makes them unsuitable for poultry feeding.

WHEY

Whey arises from the manufacture of cheese and may be available either condensed as whey paste or as dried whey meal.

Whey paste is more suitable for use in wet mashes.

Millet (*Setaris sp.*)

Rate of inclusion:

Youngstock	5 to 20 per cent
Adults	20 to 40 per cent

Millet has a somewhat similar feeding value to barley and oats. The seeds are rather small, and are sometimes included with maize grits and pin-head oatmeal in grain mixtures designed for chicks. There are many varieties.

Molasses

Rate of inclusion:

Youngstock	1 to 2 per cent
Adults	2 to 3 per cent

Molasses is a liquid by-product from the manufacture of sugar, and both cane and beet molasses are available. It is a useful source of energy and contains 49 to 52 per cent sugar and 20 to 25 per cent water. Cane molasses is usually higher in sugar than that made from beet. In addition it is a rich source of minerals, with 3 to 4 per cent potassium as well as calcium, and smaller amounts of sodium and trace minerals.

The most popular method of using the food is at a level of 2 to 3 per cent as a binding agent in the manufacture of pellets. However, molasses tends to have a laxative effect and for this reason should not generally be used at more than 2·5 per cent for birds on litter.

Molasses is useful in improving the palatability of rations and also helps to reduce dustiness.

The main problem in the use of molasses is that of handling it. Molasses is a sticky syrupy liquid that can only properly be used in modern mills which have been specially equipped for the purpose, including bulk methods of handling.

Molasses is also used as a substrate for the preparation of yeast protein.

Mowrah (*Bassia latifolia*)

Extracted mowrah-seed meal is poisonous and is not suitable for use as a feed ingredient.

Mustard-seed meal

See Rapeseed meal.

Niger (*Guizotia abyssinica*)

Extracted niger-seed meal is not a suitable poultry food and is somewhat similar in feeding value to undecorticated groundnut meal.

The seed is small and black in colour, and is grown in tropical areas.

Oats (*Avena sativa*)

Rate of inclusion:

Youngstock	2·5 to 10 per cent
Adults	10 to 30 per cent

Ground oats, owing to their high fibre content, are not recommended for use in broiler mixtures but nevertheless oats are a valuable poultry food.

Growing turkeys find whole oats very palatable and, given the opportunity, will eat large quantities; for this reason oats are valued by some turkey farmers.

The term 'groat' is applied to the kernels after the removal of the hulls and, on grinding the groats, the meal is referred to as 'oatmeal'. All forms of groats are valuable for poultry, particularly for youngstock and may be included in mixtures at the rate of 60 per cent.

Oatmeal is usually too expensive to use for poultry feeding, but is particularly valuable for inclusion in starter foods.

Pinhead oatmeal is often mixed with maize grits and millet in mixtures for young chicks.

Oat feed, a by-product that arises from the preparation of oats for human use, is generally of such high fibre content that it is quite unsuitable for use by poultry.

The term Sussex Ground Oats was at one time used to apply to oats that have been ground in a stone mill. Whole oats may be passed through rollers to produce 'bruised' or 'rolled' oats and in this form young chickens are able to pick out the kernels.

Oyster shell

See Calcium and Phosphorus.

Palm-kernel meal (*Elaeis guineenis*)

Rate of inclusion:

Growers and adults	5 to 10 per cent

Palm-kernel meal is the ground residue following the extraction of the oil from the fruit of the African oil palm. It is not a good-quality poultry food ingredient since it is high in fibre and is only a medium protein-rich feedstuff. It should always be tested for aflatoxin before use.

Palm flour has been included in chick foods at a level of 17·5 per cent as a complete replacement of rice polishings, apparently with success.

Palm oil is extracted from the pericarp of the palm nut and palm-kernel oil from the inner kernel. The meal is prepared from the residues of the palm kernels following the removal of the oil either by pressing or by solvent extraction.

Extracted babassu palm oil meal is a poor-quality food ingredient made from the nuts of the wild palm of Brazil, *Orbignia speciosa*. The meal is not suitable for poultry but may be used for ruminants.

Peas

Field peas (*Pisum sativum*); chick peas (*Cicer arietinum*); cow peas (*Vigna unguiculata*).

Rate of inclusion:

Youngstock	2·5 to 10 per cent
Adults	5 to 20 per cent

Peas, including chick peas and cow peas, are a useful protein feedstuff of somewhat similar value to beans. The protein content ranges from 17 to 25 per cent and, compared with soya-bean meal, is low in methionine.

Whole peas are not very satisfactory for use and are best fed cracked or ground; turkeys, however, will accept whole peas. Newly harvested peas are said to give digestive upsets.

Black Indian peas may be contaminated with Mutter peas (*Lathyrus sativus*), which may be actively poisonous or, at best, sufficiently deficient in methionine to give a markedly retarded growth rate.

Pecans (*Carya illinoensis*)

Extracted pecan oil meal may become available from time to time.

The meal is not suitable for poultry feeding but may have some use for ruminants.

Phosphorus

See Calcium and Phosphorus

Poppy-seed meal (*Papaver somniferum*)

Poppy-seed meal may occasionally be offered for use but is not a suitable food ingredient for poultry. Opium may be present in the meal.

Potatoes (*Solanum tuberosum*)

Rate of inclusion:

Dried potato meal	Youngstock	5 to 10 per cent
	Adults	10 to 20 per cent

Cooked potatoes and potato meal are highly digestible carbohydrate foods that may be allowed to form up to 40 per cent of the dry matter of the total diet.

Cooked potatoes generally contain about 75 per cent moisture so that 1·8 kg (4 lb) of cooked potatoes may be regarded as equal to 0·45 kg (1 lb) of mash. Owing to their bulkiness, the daily consumption of potatoes should not exceed 160 g in the case of chickens and 220 g in the case of turkeys. Raw potatoes are less digestible and should always be washed, peeled and cooked before use.

Fresh potatoes will turn green when exposed to light and become poisonous due to the presence of the toxin solanin in the skin. Green potatoes may be used after boiling in water when the toxin is dissolved. The water must be thrown away.

Potato protein may be available following the development of a new process for the preparation of potato starch. In the new process much less water is used, thus the level of protein material in the waste water is raised to a level at which its recovery may be practical.

Poultry by-products

Rate of inclusion:

	Youngstock	*Adults*
Hydrolized feather meal Offal and blood meal Feather, offal and blood meal	2%–5%	5%–10%
Dehydrated poultry waste	—	5%–10%

BROILER-PROCESSING-PLANT WASTE

The regular processing of large numbers of broilers that takes place in modern processing plants allows the opportunity for the production of protein-rich meals of uniform quality. The whole of the waste products from the processing can be treated together or the feathers and the offals can be separated.

The yield of feathers, offals and blood represents about 11 per cent of the total liveweight of the poultry slaughtered, 6 per cent being feathers and 5 per cent offal and blood.

The proteins of feathers are known as 'keratins', and untreated they are of no value for poultry feeding. Cooking under steam pressure changes the protein into a useful form: hydrolized feather meal. The process is best carried out at 50-lb steam pressure for sixty minutes.

The protein content of feather meal may be in excess of 80 per cent but is somewhat low in lysine, methionine and tryptophan. When the whole waste from the plant is processed together the protein range is 55 to 70 per cent.

DEHYDRATED POULTRY WASTE (D.P.W.)

The processing of manure from birds housed in laying cages, to produce a pathogen-free dehydrated animal protein food, is now being undertaken commercially in many countries.

D.P.W. can be a useful source of supplementary protein. The crude protein estimate may be as high as 36 per cent but half of this is non-protein nitrogen which cannot be utilized by poultry. The fibre content averages about 2 per cent. The energy level is low but this may be overcome by the use of fat.

One of the main criticisms in the past against the use of D.P.W. has been the suggestion that the product could act as a carrier of pathogenic organisms, particularly salmonellae. The processing plant must, of course, be of a high standard with special arrangements to prevent contamination.

Poultry manure produced from poultry not housed in laying cages may not be suitable owing to the possibility of recycling harmful drugs. By contrast it is not customary to include drugs in the diet of caged laying birds and the making of the D.P.W. from their droppings becomes a practical proposition.

Pulses (*See also* Peas)

The term pulses is applied to the fruits of various leguminous plants that are suitable for animal feeding. As a group the pulses contain in the region of 20 per cent protein.

Pulses contain rather more lysine than cereals but are usually lower in methionine content. The more easily available pulses are discussed under their individual common names.

Table 47 (*Palmer-Jones and Halliday, 1971*)

Common name	*Latin name*
Pigeon pea	*Cajanus cajan*
Chick pea, Bengal gram	*Cicer arietinum*
Horse gram	*Dolichos uniflorus*
Hyacinth bean	*Lablab niger*
Lathyrus pea, Khesari dhal	*Lathyrus sativus*
Lentil, split peas	*Lens esculenta*
Lupin, South American	*Lupinus sp.*
Lupins, yellow	*Lupinus sp.*
Lima bean, butter bean	*Phaseolus lunatus*
Mung bean:	
green	*Phaseolus aureus*
black	*Phaseolus mungo*
Kidney beans	*Phaseolus vulgaris*
Peas	*Pisum sativum*
Velvet bean	*Stizolobium deeringianum*
Broad bean	*Vicia faba*
Cow pea	*Vigna unguiculata*
Bambara groundnut	*Voandzeia subterranea*

Rapeseed meal (extracted)

Rape (*Brassica napus*); field mustard (*Brassica campestris*); mustard (*Sinapis sp.*).

Rate of inclusion:

Youngstock	2 to 5 per cent
Adults	5 to 10 per cent

Rapeseed meal may not be a satisfactory ingredient for poultry foods and its use requires considerable care. Though there is a long history of successful use in Canada especially with birds laying white-shelled eggs, experience with brown-shelled strains suggests that rapeseed may produce serious taints. The meal may result from rape, field mustard or mustard seeds. Rape is also known as cole and coleseed.

Rapeseed might contain goitrogens and the use of the meal may cause perosis in young chickens with reduced egg production and tainted eggs. Non-goitrogenic varieties are under widespread investigation and the new varieties tower and ergus are now available for use.

The meal may have a protein content of 40 per cent and fibre ranges from 10 to 13 per cent. The yield of meal from the original seed is in the region of 50 per cent.

Rice meal (polished) (*Oryza sativa*)

Rate of inclusion:

Youngstock	5 to 20 per cent
Adults	10 to 40 per cent

Whole rice, also called rough rice, whether fed whole or ground, is not a very suitable ingredient for poultry feeding because of fibrous husks which make the rice unpalatable. Whole rice may be contaminated with silica.

Once the husks have been removed the rice becomes 'brown rice'; this is prepared for human use by removing the bran, and is then known as white rice. White rice is usually further processed or polished and the residues are called rice polishings.

The samples of rice that become available for poultry feeding are usually those that have been rejected for human use and consist of white polished rice. Whether whole or ground, this is a very palatable and extremely digestible carbohydrate food for poultry of all ages. Whole grains are readily consumed by growing and adult birds but should be cut if fed to young chicks.

Rice by-products

RICE BRAN AND RICE POLISHINGS

Rate of inclusion:

Growing and adult stock 5 to 20 per cent

Rice by-products are frequently unsuitable for use as ingredients for starter foods owing to their extremely variable nature. Only freshly produced samples should be considered. Because of their high oil content rice polishings may rapidly become rancid.

Rice bran

Rice bran meal, also known as rice meal, may contain 12 per cent protein and from 12 to 18 per cent oil. The meal may also be a useful source of some of the vitamins of the B group. Good-quality material may be used as a source of energy in mixtures for growing and adult poultry.

Rice polishings

The feeding value of rice polishings depends upon the degree of polishing to which the grain has been subjected. It is unlikely to contain much of the rice flour.

Roots

Rate of inclusion:

Up to 2 kg (5 lb) per 100 birds, daily

Fodder beet, mangolds and swedes although having little nutritive value are sometimes useful for birds kept under intensive conditions where cannibalism is a problem. The roots are best split so that the birds can get at the succulent content.

The chief value of roots lies in the fact that they provide the birds with something to do and, therefore, may help to prevent boredom and the subsequent vices of feather picking and cannibalism.

Rye (*Secale cereale*)

Rate of inclusion:

Growing and adult stock 5 to 15 per cent

Rye meal is not an ingredient of choice for poultry feeding. Poultry generally find rye somewhat unpalatable although good-quality grain has a nutritional value similar to that of wheat.

Rye is frequently contaminated with the fungus ergot and, when this is the case, must not be fed to poultry.

Safflower (*Carthamus tinctorius*)

Rate of inclusion:

Youngstock 5 to 10 per cent

Adults 10 to 20 per cent

Extracted decorticated safflower meal may be used as a supplementary source of protein. The protein is not, however, of high quality and requires supplementing with the amino acids lysine and methionine.

Safflower is also known as kurdee.

Sago-pith meal (*Metroxylon sagus; Metroxylon rumphii Mart*)

Rate of inclusion:

Growers and adults 10 to 30 per cent

Sago-pith meal is produced from the pith of the sago palm. It is a useful source of energy but is low in protein and has a somewhat similar value to that of cassava meal.

Seaweed meal (*Chlorophycae* (green); *Rhodophycae* (red))

Seaweed meal is a rich source of minerals and vitamin B_{12}. It may be used at a level of 2 per cent in growers' rations. The use of seaweed meal has been associated with discoloured yolks when included in rations for laying birds.

Sesame meal (*Sesamum indicum; Sesamum radiatum; Sesamum orientale*)

Rate of inclusion:

Youngstock	5 to 10 per cent
Adults	10 to 20 per cent

Extracted sesame meal is a good source of vegetable protein. The protein content may vary from 40 to 48 per cent.

The meal is rather low in the amino acids lysine and methionine, and these should be provided from other sources when sesame meal is included in starter foods.

Sesame is also known as benne, gingili and teel.

Shea nut (*Butyrospermum parki*)

Shea nut has a bitter taste and is unsuitable for poultry feeding.

Single-cell protein (*S.C.P.*)

Rate of inclusion:

	Youngstock and adults:
Bacterial methanol and methane protein Yeast petroleum protein	0 to 10 per cent

The terms 'single cell', 'fermentation' and 'microbial' protein are used to describe protein-rich products of fermentation making use of a variety of micro-organisms. Bacterial and yeast S.C.P. are now commercially available and are first-quality products of high biological value.

Bacterial and yeast S.C.P. have been equated with the best-quality fish meal in nutritional value and have an excellent amino-acid balance. Yeast S.C.P. however has been shown to require supplementation with methionine.

The protein content of S.C.P. may range from 30 to over 80 per cent according to the micro-organism and substrate used. Experimental work in many parts of the world has demonstrated the acceptability of bacterial and yeast S.C.P. in the diets of poultry.

The technology concerning the commercial production of S.C.P. is well established and plants are being developed in many countries. A plant may produce 100,000 tonnes or more a year independent of local climatic conditions. The micro-organisms are those that will use the carbon in certain compounds, known as feedstocks or substrates, and will multiply rapidly with a new generation every hour. It is the micro-organism itself that becomes the protein source and S.C.P. is the end product of the process.

An advantage of S.C.P. over other animal protein foods is the unlikelihood that it will be a carrier of pathogenic organisms.

A wide range of micro-organisms and substrates may be used for the production of S.C.P., including those listed in Table 48.

Table 48 The more common processes of single-cell protein

Micro-organism	*Feedstock*	*Trade name*
Bacteria	Methanol	Pruteen
Bacteria	Methane	Shell
Fungi	Sulphate spent liquor from wood pulp mills	Pekilo
Fungi	Molasses, starch, carobs	—
Fungi	Farm livestock waste	—
Yeasts	Petroleum products, gas oil	Toprina
Algae	Chlorella sp.	—

Other feedstocks that may be employed include dates, potatoes, whey and urban waste.

Soapstock

See Vegetable-oil seed and refining products.

Sorghum (*Sorghum vulgare*)

Rate of inclusion:

Youngstock	10 to 40 per cent
Adults	20 to 50 per cent

Sorghum grains include a number of varieties such as dari, durra, feterita, guinea corn, hagari, kaffir corn and milo, whose seeds are rich in carbohydrates.

The composition of good-quality sorghum is similar to wheat, but sorghums are very variable in quality. Some varieties are rejected by birds and this is probably due to their bitter or astringent taste as a result of having a higher than average tannin content. The bird-resistant varieties are usually brown coated, have high yields and are resistant to disease, head moulds and weather damage.

The use of the less satisfactory sorghums may result in mottled egg yolks, poorer food conversion and lower weight gain in chicks and in an increased incidence and severity of a leg deformity. Growth may be depressed when sorghum contains 1·6 per cent or more of tannin, but harmful results may be mitigated by the use of supplementary methionine or removal of the pericarp from the resistant grains.

Soya beans (*Glysine max.*)

Rate of inclusion:

	Mashes	*Pellets*
Full fat soya-bean meal	10%–25%	10%–40%
Ext soya-bean meal	10%–40%	10%–40%

Extracted soya-bean meal has an average protein content of 44 per cent and the protein is the best balanced in respect of amino acids of all the normal vegetable-protein sources. It is a good source of choline and glycine but, compared with animal protein-rich foods, it is low in methionine.

With proper amino-acid supplementation, soya-bean meal may provide the whole of the protein ingredient in mixtures for poultry of all species.

Only good-quality soya bean meal should be used; the best samples have a light brown colour and a nut-like flavour.

Soya-bean meal is made from the residues of whole beans following extraction of the oil. The yield of soya-bean meal from the whole beans amounts to about 60 per cent.

In their raw state soya beans are unsuitable for animal feeding because of the presence of growth-inhibiting factors. However, these factors can be overcome by proper processing, which is referred to as 'cooking' or 'roasting'.

The growth-inhibiting factors include a toxic protein haemglutinin called soyin, which depresses appetite, and a tryptsin inhibitor which reduces the availability of methionine and cystine.

Undercooking or overcooking may result in the production of

a less satisfactory meal. An indication of the degree of cooking may be obtained by the phthalein test (cresol red) or heat test. An acceptable range is 3·5 to 4·3.

Table 49 Soya processing quality tests (*Colborn International*)

DEGREE OF HEATING	DYE ABSORPTION *Mg of cresol red absorbed per gram of meal*	UREASE DETERMINATION *Mg of urea destroyed per gram of meal*	TRYPSIN INHIBITOR *Mg trypsin per gram of meal*
Raw or much under-heated	2·0–3·0	900–580	50–40
Underheated	3·1–3·4	580–500	40–20
Slightly underheated	3·5–3·7	500–200	20–1
Properly heated	3·8–4·3	200–55	Less than 1
Overheated	Over 4·3	50–10	

Slight underheating is usually considered unimportant but whenever possible the soya-bean meal should be within the properly heated range.

Suggested specification:

Moisture	12%
Protein	44%
Oil	1%
Fibre	5%

FULL-FAT SOYA BEANS

Whole soya beans, following cooking and grinding, may be used for poultry feeding and are referred to as roasted full-fat soya-bean meal.

The digestibility of the fat in the soya bean can apparently be enhanced by finely grinding the meal.

Steamed bone flour

See Calcium and Phosphorus.

Sugar-beet pulp (*Beta saccharifera*)

Dried sugar-beet pulp is a by-product obtained during the manufacture of sugar. It is extremely high in fibre, is of low biological value and is not a satisfactory ingredient for poultry feeding.

Sunflower-seed meal (*Helianthus annus*)

Rate of inclusion:
Growing and adult stock 5 to 10 per cent

Good-quality ground decorticated sunflower-seed meal is a useful source of supplementary protein. It has a protein content ranging from 32 to 46 per cent according to the method of preparation. The protein is deficient in lysine but is a useful source of niacin. Samples high in fibre should be avoided.

Undecorticated sunflower-seed meal is not a suitable poultry food ingredient.

The use of sunflower-seed meal has been associated with the appearance of characteristic egg-shell stains which develop after laying.

Tallow

See Fats and Oils.

Tapicoca meal

See Cassava.

Triticale

Rate of inclusion:
Youngstock and adults 20 to 60 per cent

Triticale is a new species of cereal developed from durum wheat and rye that is still in the experimental stage.

The energy value of triticale is said to be similar to that of wheat, but the protein content may range from 14 to over 18 per cent.

Experimental work has shown that triticale may be used in place of wheat in poultry foods.

Urea

Urea is not suitable for inclusion in poultry food mixtures since birds are not able to make use of urea nitrogen.

Vegetable-oil seed and refining products

Rate of inclusion:

	Youngstock and adults:
Oil-seed meals, extracted and decorticated	5%–30%
Acidulated soapstock (not finisher foods)	1%–2%
Dried acidulated soapstock	1%–2%

The residues following the production and refining of vegetable oils can provide valuable animal food ingredients in the form of extracted oil-seed meals and soapstock.

EXTRACTED OIL-SEED MEALS

Extracted oil-seed meals are the residues that remain following the extraction of the vegetable oil from a variety of fruits and seeds of plants, e.g. soya. Soapstock may sometimes be included with the oil-seed residues.

The nutritive value of the residues varies considerably according to the species of plant involved, the manner in which the fruit or seed is processed prior to the extraction of the oil and according to the method of oil extraction. The protein content may range from 20 to 50 per cent.

The outer coating of many of the oil seeds is extremely fibrous and, if this is included in the residues, the products will be of poor nutritive value and often unsuitable as a ration ingredient. The process of removing the outer coating is known as decortication. Only decorticated oil-seed residues should be considered for poultry feeding.

Oil is extracted from the seeds by two methods known as the press and the solvent methods. The solvent method is the most efficient and involves dissolving the product in a solvent, the resulting residues being dried and ground.

There are two different ways of pressing the oil from the seeds. The oldest method is the hydraulic press. The second method is the expeller screw and the meals resulting are sometimes called expellers.

The yield of animal food that results following the extraction of the oil varies considerably. In the case of soya beans the yield may equal 60 per cent of the original beans whereas with cotton seed the yield may be about 30 per cent.

SOAPSTOCK

Soapstock is the liquid that results from the alkali process of oil refining, e.g. soya-oil soapstock and cotton-seed soapstock.

Acidulated soapstock from a variety of oil seeds is a useful source of energy and is also valuable as a means of reducing dustiness in food mixtures. The product should have a peroxide value of not more than 10 milli-equivalents per kg and should contain an antioxidant. The fatty acid content should be a minimum of 85 per cent, unsaponifiable matter 6 per cent maximum and a maximum of 1 per cent insoluble matter.

It should be tested for chlorinated pesticide residues, aflatoxin and, if derived from cotton seed, for gossypol.

Acidulated soapstock should be withdrawn from food mixtures for table birds four weeks prior to their being processed because the use of soapstock may result in soft and unsatisfactory carcase fat.

Dried soapstock meal is a useful source of energy and appears to be a natural antioxidant and pigmenter.

Vetch (*Vicia sativa*)

Rate of inclusion:
Youngstock 5 to 10 per cent
Adults 10 to 20 per cent

Ground-vetch seeds have a similar feeding value to beans. The protein content is usually in the region of 26 per cent, but the balance of amino acids is poor.

Walnuts (*Juglans regia*)

Poor-quality walnuts and extracted walnut-oil meal may occasionally be available for animal feeding. The meal may well contain broken shells and is not suitable for poultry.

Whale meal

Rate of inclusion:

Youngstock and adults 2·5 to 10 per cent

Good-quality whale-meat meal may be used to supply the entire protein requirements for poultry. Whale-meat meal has a high biological value, but samples of more than 10 per cent oil should be avoided.

Whale-meat meal is sometimes blended with ground whale bones and this reduces the protein content to something approaching that of fish meal.

Wheat (*Triticum sp.*)

Rate of inclusion:

Young and adult stock 20 to 60 per cent

Wheat, either whole or ground, is an excellent grain for poultry and can form the sole source of cereals for growing and adult poultry.

Wheat should be coarsely ground for inclusion in food mixtures owing to the tendency of the flour formed in grinding to make the mixture unpalatable and leading to beak impaction.

The use of new-season wheat is not recommended and has been associated with outbreaks of pullet disease.

Research work has indicated that wheat protein is limited in lysine, threonine and valine.

Emmer grain, a primitive variety of wheat, is sometimes available in limited quantities and may be used at the rate of up to 25 per cent in non-specialist rations. The nutritional value of Emmer grain is inferior to that of wheat, the protein value is similar but the fibre level may be 10 per cent. Little information is available concerning the amino-acid and vitamin value of Emmer grain.

Wheat by-products

WHEAT BRAN

Rate of inclusion:

Youngstock 5 to 10 per cent

Adults 10 to 25 per cent

Wheat bran consists of the outer coating of the wheat grain and tends to be rather high in fibre and is, therefore, unsuitable for use in high-efficiency food mixtures.

Bran is very light in weight and supplies bulk and, for this reason, is useful in lightening dense food mixtures. It readily absorbs water, and is useful when a wet-mash feeding system is followed.

The protein content of bran ranges from 10 to 12 per cent and the fibre content from 9 to 11 per cent.

WHEAT MIDDLINGS

Rate of inclusion:

Youngstock	10 to 20 per cent
Adults	20 to 40 per cent

Wheat middlings, also known as wheatfeed, consist of the fine particles of the outer coating of the wheat kernel and also particles of the germ.

The protein content of middlings ranges from 12 to 14 per cent and quality depends upon the flour extraction rate. Fibre levels range from 4 to 8 per cent.

The fibre content of middlings usually precludes its use in high-efficiency mixtures but, nevertheless, middlings is a valuable poultry food.

On all occasions only freshly prepared middlings should be used.

Suggested specification:

Moisture	10 per cent
Protein	18 per cent
Fibre	6 maximum
Acid value of the oil	30 maximum

WHEAT-GERM MEAL

Wheat-germ meal usually contains about 25 per cent protein and is a valuable poultry food when cooked and may be used at a level of 2·5 per cent. It is, however, usually too expensive to use in poultry mixtures.

Wheat-germ meal is rich in oil and vitamin E, and will rapidly become rancid if improperly stored.

Whey

See Milk.

Yeast (*Saccharomyces cervisiae*)

Rate of inclusion: Unextracted dried yeast
Young and adult stock 2·5 to 5 per cent

Fresh and dried brewers' yeast and unextracted dried yeast are extremely valuable sources of protein, minerals and some important vitamins, particularly those of the B complex. Yeast may also contain an unidentified nutrient factor.

Wet yeast ferments rapidly and, for this reason, should not be stored. It may, however, be used fresh as a part of wet mash at a level of up to 10 per cent.

On drying, wet yeast forms into large flakes which must be ground before inclusion in poultry foods.

YEAST PROTEIN

See Single-cell protein.

TORULA YEAST

Torula yeast (*Torulopsis utilis*), produced by fermentation processes using molasses or dates as the base, may be a useful source of supplementary protein fed at up to 5 per cent of the total diet.

The crude-protein content of torula may range from 45 to 50 per cent and the product may be low in methionine.

Little scientific work has been carried out concerning the nutritional role of torula yeast in specialist poultry food mixtures.

Torula yeast is rather bulky and in one experiment the food intake was improved when the meal containing the torula yeast was pelleted.

APPENDIX TWO

Analysis of selected ingredients

The analytical details given in Table 50 have been prepared for guidance when formulating under field conditions. The tables have been collated from many sources and, in the case of the amino acids and vitamins, the figures have generally been the lower rather than the higher of the sources consulted. In all cases the aim should be to establish, by the usual methods, the nutritional values of the actual ingredients used (see pages 219–234).

The feedstuffs included in Table 50 are far fewer than the very large number discussed in the first part of this chapter. They do, however, include the ingredients most commonly used in many parts of the world and the analytical information is reasonably complete. Other ingredients, when of special local importance, should be analysed in a similar way and then compared with the most suitable standard ingredient before substitution is attempted.

Table 50 Typical analysis of selected poultry foods

Ingredient	*Dry matter*	*ME kcal/kg*	*Protein per cent*	*Fat per cent*	*Fibre per cent*
Alfalfa meal	93	1,720	20	3·5	21·0
Barley	89	2,650	11·5	1·5	7·0
Beans	90	2,350	21	1·0	4·5
Blood meal	91	3,100	80	0·8	1·0
Coconut meal ext.	93	1,650	20	6·0	13·0
Cottonseed meal ext.	92	2,600	41	0·5	12·0
Feather meal	95	2,600	85	2·5	1·5
Fish meal	91	2,800	60	4·0	1·0
Groundnut meal ext.	92	2,500	45	1·0	11·0
Maize	86	3,200	8·8	3·8	2·0
Maize-gluten meal	91	2,500	42	2·0	4·0
Meat meal	94	2,000	55	7·1	2·5
Meat and bone meal	94	1,900	45	8·6	2·1
Oats	89	2,500	11·5	4·0	11·0
Safflower meal ext.	93	2,650	46	1·3	5·0
Sesame meal ext.	93	2,570	46	5·0	6·0
Sorghum	88	3,100	10	3·7	2·0
Soya-bean meal ext.	89	2,200	44	0·5	6·0
Sunflower meal ext.	92	2,100	45	2·5	11·0
Wheat	89	2,900	12	2·0	3·0
Wheat bran	89	2,490	14·5	3·8	10·0
Wheat middlings	89	2,000	15·0	4·0	7·0

Table 50 (*continued*)

Typical analysis of selected poultry foods

Ingredient	MINERALS Ash per cent	Calcium per cent	Available Phosphorus per cent	Manganese mg/kg
Alfalfa meal	10·5	1·45	0·11	30
Barley	2·6	0·08	0·15	15
Beans	6·0	0·11	0·21	13
Blood meal	4·5	0·30	0·22	1
Coconut meal ext.	6·5	0·21	0·22	55
Cottonseed meal ext.	6·5	0·16	0·48	21
Feather meal	3·8	0·20	0·75	5
Fish meal	22·0	6·0	3·0	20
Groundnut meal ext.	4·8	0·20	0·24	30
Maize	1·5	0·03	0·10	5
Maize-gluten meal	4·5	0·16	0·16	7
Meat meal	32·0	7·4	3·8	9
Meat and bone meal	38·0	10·0	5·1	9
Oats	3·1	0·10	0·14	38
Safflower meal ext.	7·8	0·40	0·50	48
Sesame meal ext.	11·5	2·0	0·60	48
Sorghum	1·8	0·04	0·13	14
Soya-bean meal ext.	6·0	0·32	0·26	27
Sunflower meal ext.	6·6	0·38	0·37	25
Wheat	1·6	0·04	0·20	78
Wheat bran	5·5	0·14	0·20	84
Wheat middlings	4·9	0·09	0·34	122

Table 50 (*continued*)

Typical analysis of selected poultry foods

	AMINO ACIDS					
Ingredient	*Arginine per cent*	*Lysine per cent*	*Methio-nine per cent*	*Cystine per cent*	*Trypto-phan per cent*	*Glycine per cent*
Alfalfa meal	0·9	0·87	0·30	0·23	0·46	1·0
Barley	0·5	0·35	0·10	0·19	0·13	0·30
Beans	1·0	1·40	0·17	0·10	0·24	0·8
Blood meal	1·7	1·40	1·00	1·40	1·05	1·15
Coconut meal ext.	1·9	0·48	0·31	0·24	0·18	0·91
Cottonseed meal ext.	2·7	0·65	0·29	0·30	0·20	1·00
Feather meal	5·6	1·55	0·50	3·00	0·50	6·80
Fish meal	3·0	4·50	1·65	0·52	0·62	4·20
Groundnut meal ext.	4·5	1·60	0·40	0·60	0·45	2·10
Maize	0·4	0·18	0·09	0·09	0·07	0·35
Maize-gluten meal	1·4	0·80	1·00	0·60	0·20	1·50
Meat meal	3·7	2·57	0·75	0·68	0·34	6·30
Meat and bonemeal	3·0	2·20	0·53	0·30	0·20	6·01
Oats	0·7	0·32	0·18	0·18	0·16	0·50
Safflower meal ext.	1·9	0·70	0·33	0·35	0·26	1·10
Sesame meal ext.	4·2	1·20	1·25	0·58	0·60	3·90
Sorghum	0·3	0·20	0·13	0·10	0·09	0·30
Soya-bean meal ext.	3·2	3·00	0·65	0·67	0·63	2·30
Sunflower meal ext.	3·3	1·70	0·80	0·70	0·50	1·70
Wheat	0·6	0·35	0·18	0·17	0·17	0·70
Wheat bran	0·9	0·58	0·10	0·30	0·27	0·90
Wheat middlings	0·7	0·45	0·18	0·18	0·23	0·89

Table 50 (*continued*)

Typical analysis of selected poultry foods

Ingredient	VITAMINS *mg/kg* *Choline*	*Niacin*	*Pantothenic acid*	*Riboflavin*
Alfalfa meal	1,500	44·0	20·0	10·6
Barley	1,030	52·9	6·2	1·5
Beans	1,760	24·2	3·1	1·9
Blood meal	750	30·8	2·2	1·4
Coconut meal ext.	920	24·2	6·17	3·1
Cottonseed meal ext.	2,650	35·5	7·1	4·0
Feather meal	2,000	36·8	16·9	4·8
Fish meal	1,550	55·0	6·6	6·0
Groundnut meal ext.	2,000	165·0	50·0	4·6
Maize	530	20·0	4·8	1·1
Maize-gluten meal	330	44·0	9·9	1·5
Meat meal	2,000	57·0	4·8	5·0
Meat and bone meal	1,320	35·0	4·6	3·7
Oats	920	13·0	12·0	1·1
Safflower meat ext.	3,250	22·0	39·6	2·4
Sesame meal ext.	1,500	13·0	5·5	3·3
Sorghum	675	42·0	11·4	1·2
Soya-bean meal ext.	2,640	26·5	13·2	2·8
Sunflower meal ext.	2,900	220·0	10·0	3·0
Wheat	870	46·0	11·0	1·1
Wheat bran	988	198·0	26·0	3·0
Wheat middlings	1,100	52·0	13·5	1·5

APPENDIX THREE

Conversion tables

Table 51 Mineral conversions

	Multiply by	*Reciprocal, multiply by*
Calcium to calcium oxide	1·4	0·714
Calcium to calcium carbonate	2·5	0·4
Calcium oxide to calcium carbonate	1·8	0·56
Chlorine to sodium chloride	1·65	0·607
Manganese to manganese sulphate	2·75	0·364
Phosphorus to phosphoric acid	2·29	0·436
Magnesium to magnesium oxide	1·658	0·603
Copper to copper sulphate	2·51	0·398
Iron to iron sulphate	2·72	0·367

Table 52 Temperature

Farenheit to Centigrade
subtract 32 and multiply by $\frac{5}{9}$

Centigrade to Fahrenheit:
Multiply by $\frac{9}{5}$ and add 32

Table 53 Metric conversions

(a) LENGTH

1 in	= 25·4 mm = 2·54 cm
12 in = 1 ft	= 30·48 cm = 0·3048 m
3 ft = 1 yd	= 91·44 cm = 0·9144 m
1,760 yd = 1 mile	= 1·609 km = 1,609·347 m
1 mm	= 0·03937 in
10 mm = 1 cm	= 0·394 in = 0·0328 ft
100 cm = 1 m	= 39·37 in = 3·281 ft = 1·094 yd
1,000 m = 1 km	= 1,093·6 yd = 0·6214 miles

Table 53 (*continued*)

(b) SQUARE MEASURE – AREA

1 sq in	= 6·452 cm^2
144 sq in = 1 sq ft	= 0·093 m^2
9 sq ft = 1 sq yd	= 0·836 m^2
4,840 sq yd = 1 acre	= 4,047 m^2 = 40·469 a = 0·405 ha
640 acres = 1 sq mile	= 1,259·0 ha = 2·590 km^2
1 cm^2	= 0·155 sq in
100 cm^2 = 1 dm^2	= 15·5 sq in
100 dm^2 = 1 m^2	= 10·765 sq ft = 1.196 sq yd
100 m^2 = 1 acre	= 119·599 sq yd
100 a = 10,000 m^2 = 1 ha	= 2·471 acres
100 ha = 1 km^2	= 0·386 sq miles

(c) CUBIC MEASURE – VOLUME

1 cu in	= 16·387 cm^3
1,728 cu in = 1 cu ft	= 28·316 dm^3 = 0·0283 m^3
27 cu ft = 1 cu yd	= 0·765 m^3
1 pint = 20 oz	= 0·578 litre
2 pt = 1 qt	= 1·137 litre
4 qt = 1 gall (Imp) = 10 lb = 0·16 cu ft	= 4·546 litre = 0·045 hl
1 gall (USA) = 0·83 gall (Imp)	= 3·785 litre = 0·038 hl
1 cm^3	= 0·061 cu in
1,000 cm^3 = 1 dm^3 = 1 litre	= 61·023 cu in = 0·035 cu ft
1,000 dm^3 = 1 m^3 = 1,000 l = 10 hl	= 35·314 cu ft = 1·308 cu yd
1 litre = 1,000 cm^3	= 1·76 pt = 0·88 qt
100 l = 1 hl	= 21·998 gall (Imp) = 26·418 gall (USA)
1 litre = 1,000 ml = 100 cl = 10 dl	

(d) WEIGHT

1 oz	= 28·35 g
16 oz = 1 lb	= 453·592 g = 0·4536 kg
112 lb = 1 cwt	= 50·80 kg
20 cwt = 1 ton (Imp) = 2,240 lb	= 1,016 kg = 1 long ton
1 ton (USA) = 2,000 lb	= 907·185 kg = 1 short ton

Table 53 (*continued*)

(d) WEIGHT

1 mcg	
1,000 mcg = 1 mg	
1,000 mg = 1 g	= 0·035 oz
1,000 g = 1 kg	= 35·3 oz = 2·2046 lb
1,000 kg = 1 metric tonne	= 0·9842 ton (Imp) = 1·102 ton (USA)

1 mcg per g = 1 ppm
1 g per ton = 1 ppm
1 Imp bushel = 8 gall = 2218·19 cu in
1 Winchester bushel (USA and Canada) = 2150·4 cu in

Table 54 Birds per square metre/square feet per bird

Sq ft per bird	*Birds per sq metre*		*Sq ft per bird*
0·5	21·5	22	0·49
0·6	17·9	20	0·51
0·7	15·3	18	0·60
0·8	13·4	16	0·67
0·9	11·9	14	0·77
1·0	10·8	12	0·90
1·5	7·2	10	1·07
1·75	6·1	8	1·34
2·0	5·4	6	1·79
2·5	4·3	4	2·69
3·0	3·6	2	5·37
3·5	3·1	1	10·75
4·0	2·7		

Table 55 Volume of ingredients

Pounds/ Cubic feet	*Pounds/ Cubic metre*	*Metric tonnes/ Cubic metre*
28	988	0·45
30	1,059	0·48
32	1,130	0·51
34	1,201	0·54
36	1,271	0·58
38	1,342	0·61
40	1,413	0·64
42	1,483	0·67
44	1,554	0·70
46	1,624	0·74
48	1,695	0·77
50	1,766	0·80
52	1,836	0·83
54	1,907	0·87
56	1,978	0·90
58	2,048	0·93
60	2,119	0·96
62	2,331	1·06

References

Page Nos	
44, 58, 61, 76	Poultry and Egg National Board, 1964, USA
90	*Feed Additive Compendium*, 1975, The Miller Publishing Co, Minneapolis, USA
104	Comben, N. 1970, Roche Products Ltd, London, UK
105	*Feedstuffs*, 27 January 1975, The Miller Publishing Co, Minneapolis, USA
141, 142, 143, 144, 182	The Agricultural Research Council, 1975, London, UK
159, 160, 161	Poultry Testing, 1975, The National Poultry Tests, Milford, UK
169–172	Pilbrow, P., 1975, private communication
174	Hubbard Farms, New Hampshire, USA
180	ADAS, Ministry of Agriculture, 1975, London, UK
180–181	ADAS and NFU Broiler Breeder Bulletin, 1975, London, UK
185	Cherry Valley Ducks, 1975, Rothwell, UK
186, 189, 191	*Poultry World*, 1975, Agricultural Press Ltd, Sutton, UK
265	Palmer-Jones, R. and Halliday, D., 1971 *The Small Scale Manufacture of Compound Animal Feed*, The Tropical Products Institute, London, UK
272	Colborn International, 1975, UK

Further reading

Books

Atschull, A. A., 1958, *Processed Plant Foodstuffs*, Academic Press Inc., New York, USA

Bolton, W., and Blair, R., 1974, *Poultry Nutrition*, HM Stationery Office, Bull. 174, London, UK

Ewing, W. R., 1963, *Poultry Nutrition*, The Ray Ewing Company, Pasadena, California, USA

Feed Additive Compendium, 1975, Miller Publishing Company, Minneapolis, USA

Feltwell, Ray, 1963, *Turkey Farming*, Faber and Faber, London, UK

Nutrient Requirements of Poultry, 1971, National Academy of Sciences, Washington, DC, USA

Nutrient Requirements of Farm Livestock – Poultry, 1975, Agricultural Research Council, London, UK

Scott, M. L., Nesheim, M. C., and Young, R. J., 1969, *Nutrition of the Chick*, M. L. Scott and Associates, Ithacha, USA

Titus, H. V. L., 1961, *The Scientific Feeding of Chickens*, Interstate, Illinois, USA

Journals

British Poultry Science, Longmans, Edinburgh, UK

Feedstuffs, Miller Publishing Company, Minneapolis, USA

Journal of Flour and Animal Feeding Milling, London, UK

Poultry Science, Michigan, USA

Poultry World, Agricultural Press Ltd, Sutton, UK

Index